HORTICULTURE: A BASIC AWARENESS

Second Edition

Robert F. Baudendistel

Reston Publishing Company, Inc., Reston Virginia
A Prentice-Hall Company

Library of Congress Cataloging in Publication Data

Baudendistel, Robert F.
    Horticulture, a basic awareness.

    Bibliography: p. 277
    Includes index.
    1. Horticulture.   2. Gardening.   I. Title.
SB318.B38   1982   635   81-17765
ISBN 0-8359-2895-0   AACR2

©1982 by
Reston Publishing Company, Inc.
A Prentice-Hall Company
Reston, Virginia 22090

10 9 8 7 6 5 4 3 2 1

Printed in the United States of America

Many thanks to the following for their assistance:

My wife and children,
My mother and aunt,
Rennie Richmond,
Wilbur Trask,
Doug Bean

Contents

Presently there are many teachers, like myself, teaching a course in horticulture in an attempt to satisfy the overwhelming interest in the growth of plants that prevails at this time. At best, the course plus a variety of copied material is the only material used by the instructor to distribute to his or her students. Also, because of the vastness of the subject, most outlines are limited to the individual instructor's own strong subject areas.

Both the public and university libraries are well stocked with numerous books dealing with the vast subject of horticulture. Some are specific in content, treating a small segment of the subject, while others provide the reader with a general wealth of knowledge in many areas of the subject. The additions to the chapters on vegetable gardens, pruning, and garden practices should make this Second Edition more helpful to its readers.

To my knowledge, there is no comparable text designed to teach an introductory course in horticulture at the community college level. Many state universities offer a variety of courses, all complemented with textbooks, but these works are far too comprehensive for a basic course.

This textbook was written with the following thoughts in mind:

1. To reach those individuals (a) considering a career in horticulture but as yet undecided about which area to specialize in, (b) interested in horticulture as an avocation, regardless of age, and (c) desiring a book containing information on many horticultural topics.
2. To acquaint the reader with the basic techniques and knowledge associated with the field of horticulture.
3. To introduce those undecided students to an employment choice not previously considered.
4. To offer to interested colleges a tailor-made, yet flexible program requiring little more than an interested instructor.
5. To provide the reader with comprehensive knowledge covering a wide

range of topics within the field of horticulture in concise, easy-to-understand language.

6. To adapt the presented material so that it can be used to present horticulture to interested adults taking night school classes.

7. To design the laboratory exercises with flexibility so that the number of plants needed can be adjusted to the amount of available space.

8. To design some of the laboratory exercises so that the students receive both a feeling of accomplishment and creativity.

9. To write the laboratory exercises with directions that are easy to follow in order to provide the reader with a high level of success.

10. To offer the reader laboratory exercises that are varied in content and designed to be both interesting and educational.

11. To coordinate the laboratory exercises with the text to enable the students the opportunity to actually perform rather than simply to be told how by the instructor.

12. To provide the reader with an appendix covering a variety of information, not found within the text, in a quick reference form.

13. To provide the reader with the metric equivalent (in parentheses) for all English weights and measures used throughout the text.

ROBERT F. BAUDENDISTEL

Introduction

Whenever the word horticulture is mentioned, most people envision the growth of plants. For some, their thoughts focus on the growth of flowers, while others immediately think of growing fruits and vegetables.

Few people, however, are totally aware that the growth and cultivation of any plant material is an exacting science requiring complete knowledge of all the factors associated with the growth of each plant and the skill necessary to manipulate the environmental conditions to insure a saleable product.

Horticulture is much more than simply the growth of plants. It encompasses knowledge about soils, their texture and reaction, pest control, watering frequency, and light requirements. On a large commercial scale, each of these growth-related items is handled by specialists.

Once the plant material has been grown successfully, it may require additional treatment by other specialists. For example, fruits and vegetables, after being harvested, must be cleaned and graded before they can be offered to the public in the fresh, canned, or frozen form.

The nurseryman grows a variety of trees and shrubs but must rely on the landscape architect to properly position these same plants in the home landscape. The floral arranger would be unable to employ his artistic talents and basic knowledge of design if he had no source of flowers.

Many horticulture-related careers enable those with artistic talents to be creative, while other careers combine a love of plant material with the desire to be a public servant. All of these careers, however, provide the individual with a sense of personal satisfaction attained by either daily or seasonal accomplishments.

On pages 269 to 270 in Appendix B is a listing of horticultural and related careers, the degree requirements for each, and addresses of where to write for additional information.

Chapter One

Basic Botanical Background

All plant enthusiasts should have a knowledge of the plant kingdom and the various terminology associated with it. In this chapter, the parts of the plant will be covered, with special attention being paid to the functions of each.

The plant kingdom is separated into four major divisions:

CLASSIFICATION OF PLANTS

1. *Thallophytes*. The simplest plant types are found in this division. They include the algae, bacteria, fungi, and lichens. Most lichens consist of an alga and a fungus living in harmony with each other.

2. *Bryophytes*. This division contains all the mosses and their relatives, the liverworts.

3. *Pteridophytes*. The ferns and their allies, the horsetails and club mosses, comprise this division.

4. *Spermatophytes*. This division comprises the largest number and the most complex of all the plants. They are referred to as the seed-bearing plants. The spermatophytes are further subdivided into the gymnosperms and the angiosperms.

The gymnosperms are characterized by being evergreen, cone-bearing plants. These include the pines, spruces, junipers, and firs (Fig. 1-1). The angiosperms are known as the flowering plants. They, too, are further subdivided into monocotyledons and dicotyledons. Monocots have leaves that are narrow with parallel veins, and their flower parts occur in sets of three or multiples of three (Fig. 1-2). Dicots are characterized by having two embryonic leaves, called cotyledons, and leaves that are broad with veins running at angles to the leaf's midrib. Their flower parts are arranged in sets of four or five (Fig. 1-3).

1

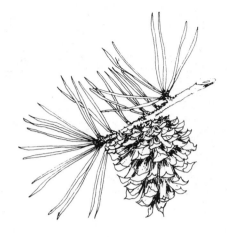

FIGURE 1-1
Pine needle and cone.

FIGURE 1-2
Monocotyledon—daffodil flower.

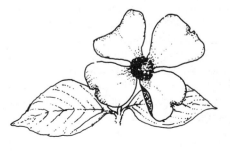

FIGURE 1-3
Dicotyledon—dogwood leaf and flower.

There are five major functions associated with most plant roots. They are

1. *Water and nutrient absorption.* This function is almost solely the responsibility of the minute root hairs. Without them, the uptake of water and nutrients would not be possible and plant death would result. This process is known as *osmosis*, or diffusion through a semipermeable membrane. This occurs whenever there is a larger concentration of water and soluble nutrients in the soil than within the plant.
2. *Water conduction.* Once the water and nutrients enter the root hairs, they are transported through the conducting tissues of the root and carried upwards through similar conducting tissues in the stem (xylem), eventually reaching their desired destination, the cells within the leaf or stem.
3. *Anchorage.* The larger roots are distributed throughout the soil area and serve to support the top growth and prevent the plant from being lifted out of the soil during periods of high winds.
4. *Storage of reserve food.* Any excess food produced by the cells in the leaves is conducted from the plant stem and may be stored in the roots for future growth.
5. *Propagation.* The roots of a few plants are used to produce new plants.

Plant stems are characterized by the production of buds, which give rise to new branches, leaves, and flowers. There are four major functions of plant stems. They are

1. *Support.* Stems are necessary as support for branches, buds, and leaves.
2. *Conduction.* Plant stems are used to conduct the water and nutrients from the roots to the leaves and back again to the roots.
3. *Storage.* The stems, like the roots, are capable of storing the excess food produced by the process of photosynthesis.
4. *Propagation.* Both herbaceous (soft, nonwoody) and ligneous (woody) plant stems can be used to obtain new plants from stem cuttings. (See Chapter 7.)

A typical plant stem has a terminal bud and many lateral buds. These buds may be either flowering buds (large and fat) or vegetative buds (narrow and thin). See Fig. 1-4. All buds arise from a *node*, and the space between individual buds is known as the *internode*. Buds may be found *opposite* each other on the stem or *alternate* on the stem, giving rise to only one leaf or stem (Fig. 1-5).

3

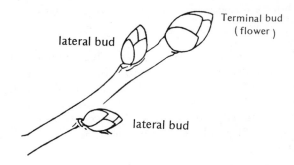

FIGURE 1-4
Plant stem—terminal flowering bud and lateral vegetative buds.

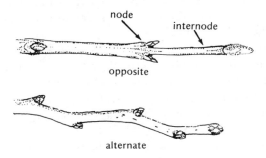

FIGURE 1-5
Plant stem—node and internode, opposite and alternate buds.

Certain plants have modified stems that may be found above ground, such as thorns, tendrils, and stolons or runners (Fig. 1-6). Modified storage stems found below ground include bulbs, corms, rhizomes, and tubers (Fig. 1-7).

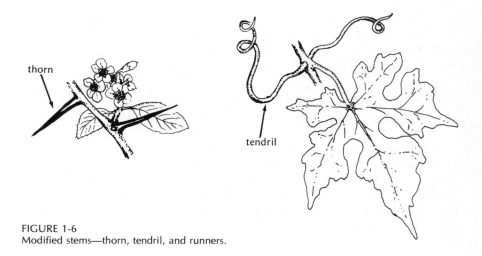

FIGURE 1-6
Modified stems—thorn, tendril, and runners.

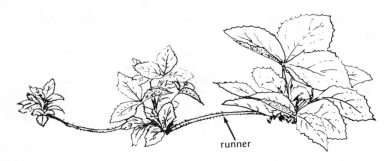

runner

FIGURE 1-6 (Continued.)

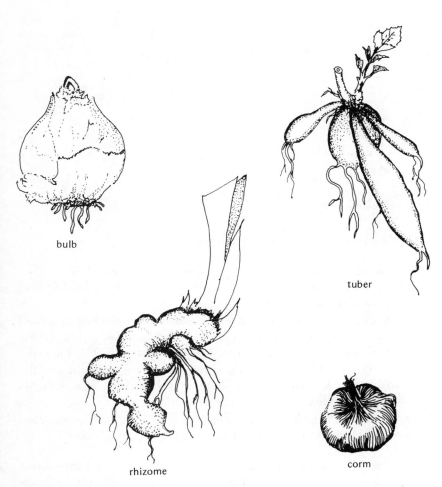

bulb

tuber

rhizome

corm

FIGURE 1-7
Storage stems—bulb, corm, rhizome, and tuber.

The Leaf   Plant leaves have three distinct parts. The main body is known as the *blade*, which is either *simple*—in one piece—or *compound*, where the leaf is divided into segments or divisions called leaflets. The *petiole*, or leaf stalk, is attached to the blade and to the leaf *base* at the other end. The base is that portion of the leaf which is attached to the plant stem (Fig. 1-8).

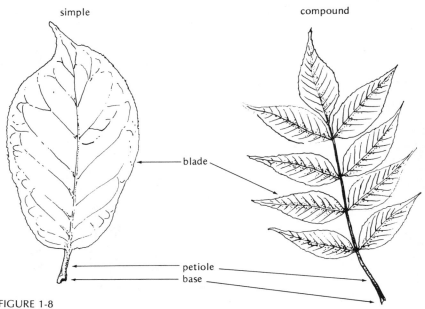

FIGURE 1-8
Parts of a leaf—simple, compound.

The four primary functions of plant leaves are

1. *Photosynthesis*. This is the unique process of leaves whereby they convert water and carbon dioxide in the presence of sunlight to food in the form of carbohydrates.
2. *Respiration*. This process involves the oxidation of the manufactured food to perform the vital life functions within the plant. This process is not unique only to leaves, since it is carried on by all living cells.
3. *Transpiration*. It is carried out and regulated by the openings in the leaves, known as *stomata*. Transpiration involves the release of excess water from the leaf's surface to the surrounding air.
4. *Propagation*. Plants having thick, fleshy leaves are usually propagated from leaf cuttings or from leaf petiole cuttings. (See Chapter 7.)

Plant leaves are most commonly used for identification and determination of an unknown plant. It is important for all horticulturalists to be familiar with the terminology used to separate one plant species from another.

Figures 1-9 through 1-12 are intended to present the more important terms associated with leaves.

lobed

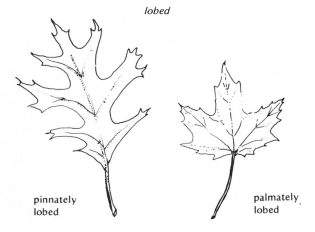

pinnately
lobed

palmately
lobed

FIGURE 1-9
Simple lobed leaves.

compound

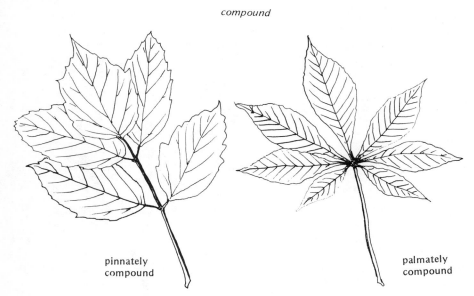

pinnately
compound

palmately
compound

FIGURE 1-10
Compound leaves.

Additional terms

Smooth. No hairs present.

Pubescent. Hairs on leaf surface.

Whorled or verticillate. Three or more leaves arise from the same node.

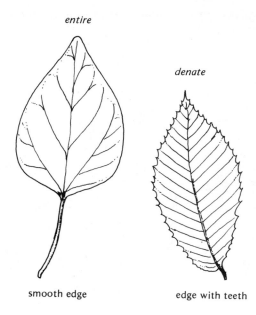

entire

denate

smooth edge   edge with teeth

FIGURE 1-11
Leaf edges.

veining

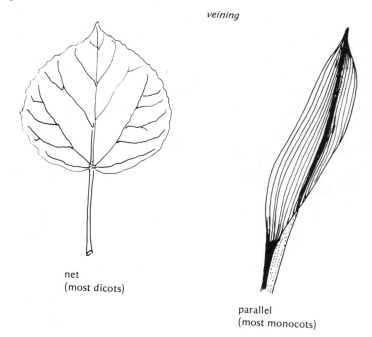

net
(most dicots)

parallel
(most monocots)

FIGURE 1-12
Leaf veining.

Certain plants have modified leaves for specific purposes. They are used for:

1. *Protection.* Leaves become bud scales and are used to protect the tissue within the bud, like the willow.
2. *Food storage.* An example is the scale leaves of the onion.
3. *Water storage.* The fleshy leaves of most succulents are able to store water.
4. *Trapping insects.* The Venus' flytrap and other insectivorous plants have leaves capable of trapping insects.

The flower is divided into two separate types of organs, referred to as *accessory* and *essential*.

The Flower

Accessory organs.   These are commonly called the *perianth*, and consist of:

1. The *calyx*, which is usually green in color and is composed of *sepals*; it is used to protect the interior flower parts.
2. The *corolla*, which is composed of *petals*. Their bright colors are used to attract the bees and insects needed for pollination and eventual seed formation. Terminology associated with the corolla includes the following:

   a. *Regular.* All the petals are similar in size and shape, as in the rose.
   b. *Irregular.* Some of the petals differ in size and appearance, as in the snapdragon.
   c. *Apopetalous.* Each petal is separate from the other.
   d. *Sympetalous.* All the petals of the flower are fused together.

Essential organs.   These are the flower parts necessary for the all-important pollination, and consist of:

1. The *stamens,* or male parts. Each stamen consists of a filament or slender stalk which is topped by an *anther* or pollen sac. Within each anther are the *pollen grains*, which are liberated at just the right moment of time.
2. The *pistil* (carpel) or female parts. Each pistil consists of a *stigma* or tip, which receives the pollen grains. The *style* is that area of the pistil between the stigma and the ovary and consists of a slender stalk down which the pollen passes. The *ovary* is the basal part of the pistil and may contain from one to many ovules or eggs, which after pollination develop into seeds.

The following terminology is used to describe and classify flowers:

1. *Perfect*. Flowers are designated as perfect when they contain both male and female organs.
2. *Imperfect*. Flowers in which either of the organs is missing are labelled as imperfect. When both the organs are found on the same plant, but in separate male and female flowers, as in the birch, they are called *monoecious*. Plants having male and female flowers on separate plants, like the Ginkgo, are labelled *dioecious*.
3. An *inferior ovary* is identified as one that appears to be united with the calyx.
4. A *superior ovary* appears above and free of the calyx.

TYPES OF
FLOWER
INFLORESCENCES

The term inflorescence is used only to describe a cluster of flowers and is related to how the flowers are arranged on the stalk. Inflorescences are of two main kinds: In the *indeterminate* inflorescence, flower buds continue to form and open as the stem grows in length, and the outer or lower flowers are older than the inner or higher ones on the stalk. The *determinate* inflorescence is characterized by having the terminal bud form a flower and halt the growth of the flower stalk. Additional flowers that do develop would be found only on side branches below the terminal flower. The diagrams in Fig. 1-13 are designed to help identify the types of inflorescences commonly encountered.

After fertilization or pollination, all the essential flower parts dry up except the ovary. It matures into a fruit or seed pod and the ovules (egg cells) become seeds. Fruits are separated into two types, *fleshy* and *dry*.

Fruits   **Fleshy types**
1. Pome: apple, pear
2. Drupe or stone fruit: cherry, plum, peach
3. Berry: grape, tomato
4. Gourd: cucumber, gourd
5. Aggregate (cluster of drupes): raspberry, blackberry
6. Multiple: fig

Dry types.   These are divided into two types:
A. Those that split open
   1. Legume: pea, bean
   2. Follicle: milkweed, larkspur
   3. Capsule or pod: poppy
B. Those that do not split open
   1. Achene: buttercup
   2. Grain: grasses, cereals
   3. Samara: maple
   4. Nut: oak, hickory

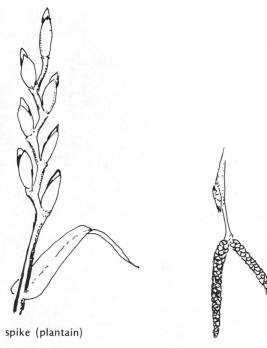

spike (plantain)

catkin (birch)

umbel (Queen Anne's Lace)

corymb (Hawthorn)

FIGURE 1-13
Types of flower inflorescences.

raceme
(lily of the valley)

panicle (lilac)

Determinate type

head (composite family)

cyme (Sweet William)

FIGURE 1-13 (Continued.)

The diagrams in Fig. 1-14 will help in the identification of the various types of fruits.

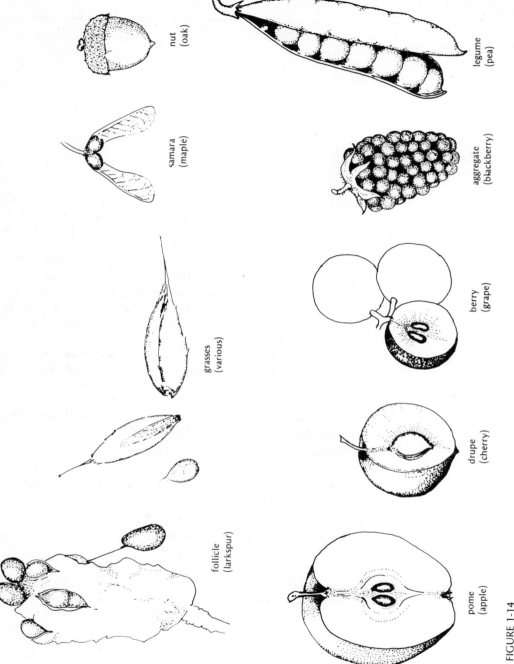

nut
(oak)

legume
(pea)

samara
(maple)

aggregate
(blackberry)

grasses
(various)

berry
(grape)

drupe
(cherry)

follicle
(larkspur)

pome
(apple)

FIGURE 1-14
Types of fruits.

13

Chapter Two
Soils

To the majority of people, soil is the material covering much of the earth's surface. Many may not realize that its formation is the product of thousands of years of decomposition and weathering, and that when it is first deposited it is unable to support much plant life.

In most cases, therefore, the grower of any plant has little choice in the selection of the soil to be used. The naturally occurring soil may be ideal, or it may limit growth because of poor drainage, lack of nutrients, or an unfavorable soil reaction. The intended purpose of every plant grower, then, is to properly adapt and continue to manage the existing soil so that it will support normal plant growth.

Even though each soil is different in some aspect from every other soil, all productive soils must be composed of five basic ingredients. They are:

1. Minerals or compounds of chemical elements obtained from the decomposition of the original rocks.
2. Organic matter or former living material that is in the process of decay.
3. Microorganisms, such as bacteria, fungi, and small animals that aid in the decomposition of the organic matter.
4. Water, which is necessary for all living things.
5. Air or oxygen, which is needed for the proper growth of plant roots.

SOIL CLASSIFICATION

Soils are classified according to the texture of their most predominate particle. For example, a sandy soil must be at least half sand. For simplicity, only the three most common soil types will be compared.

A *sandy soil* is characterized by:

1. Large visible particles (0.1 to 0.5 mm in diameter).
2. Ability to warm up quickly.
3. Ease of cultivation.

4. Inability to retain water and to prevent the loss of nutrients throughout the growing area.

The characteristics of a *clay soil* are

1. Particles invisible to the eye (less than 0.002 mm in diameter).
2. An enormous capacity to hold water.
3. An inability to dry out or warm up quickly.

A rich fertile soil, classified as a *loam*, is the most desirable soil for plants and is characterized by a mixture of sand, clay, and organic matter. Loams have good drainage, but are able to retain sufficient moisture for proper plant growth, and at the same time to provide plant roots with adequate aeration.

For increased familiarity with the existing soil, several tests should be performed by any grower before planting is started. These tests will enable the grower to correct any soil problems by making the necessary corrections or additions to the soil.

The first test that should be made on a soil is a determination of *texture*. This is easily accomplished by squeezing a moist sample of the soil between the fingers. A soil that has a predominance of sand will feel harsh and gritty, and the individual particles will be visible to the eye, feel coarse, and barely hold together. A clay soil is easily squeezed, feels very smooth and sticky, and has no visible particles. The desirable loam should feel smooth to the fingers, be somewhat sticky and feel coarse at the same time, and be easy to roll between the fingers.

The second test that should be conducted on any soil is to determine *how well it drains*. Many times, this is overlooked by plant growers, who assume that once the texture of a soil is known, its drainage potential is confirmed. All too often, we forget that the underlying soil, or subsoil, may be drastically different from the top six inches (15 cm). A simple test is to dig holes 2–3 feet (60–90 cm) in depth and fill them with water. If the water disappears within one hour, the soil displays good drainage. If water still remains in the hole one day later, drainage is poor and must be corrected, or the soil should be used only to support the growth of shallow-rooted plants.

The next two tests, involving *soil acidity* and *soil fertility*, are usually done at the same time. Good soil test kits are now available that enable anyone to test a specific soil. The procedure employed by most of these kits involves the mixing of the soil with one or more prepared solutions in a small vial. With each test, the soil and solution(s) are thoroughly shaken and then set aside until the soil has completely settled.

The eventual determination usually involves matching the color within the vial with the manufacturer's prepared color chart. The tests most

SOIL TESTING

commonly conducted involve a determination of the available amounts of nitrogen, phosphorus, and potassium, plus a determination of pH or soil acidity. It must be understood, however, that these findings are only rough approximations of the true soil picture.

To do the testing accurately, expensive sophisticated equipment is needed. Because of this, most states have soil testing laboratories for their residents who desire this service. The procedure used to obtain the sample in either case is the same. The soil sample should be taken in the early spring or late fall and should be a representative cross section of the top six inches of soil.

The test for soil acidity enables the grower to learn the pH of the soil tested. A pH of 7.0 is neutral, being neither acidic nor alkaline. Soils having pH values below 7.0 are acidic, while those above 7.0 are alkaline. The pH numbers are related to logarithms, with a base of ten; thus a soil with a pH of 5.0 is *ten times* as acidic as a soil having a pH of 6.0.

Most cultivated plants can grow in soils having a pH between 5.0 and 7.5. However, the most ideal range for the majority of plants is between 6.0 and 7.0, since within these limits all the necessary nutrients for normal plant growth are available. Acid-loving plants, such as azaleas, holly, and rhododendrons, are able to thrive in acid soils where the nitrogen is unavailable, because the mycorrhiza fungi on their roots helps change the existing ammonium nitrate into the desirable nitrate nitrogen. Most plants are unable to accomplish this feat, and when they are grown at a pH below 5.0 for an extended time they have their growth seriously retarded and eventually die.

Once the pH of the soil has been determined, the grower is able to change the existing soil condition to meet the demands of the plant material. To lower the pH of a soil, making it more acidic, aluminum sulfate is added. One and one-half (1½) lb (0.68 kg) is sufficient to lower 100 ft² (9.2 m²) of surface soil area by three-fourths of a pH value.

In highly alkaline soils, most soil organisms do not thrive too well. Certain plants may also suffer from *iron chlorosis*, a condition identified by yellow-green leaves with dark green veins, because the iron may be trapped in the soil and be thus unavailable to the plant. This chlorotic condition usually disappears as soon as the pH is lowered. The addition of a solution of chelated (meaning claw) iron to the soil will also correct chlorosis, since its chemical composition enables the iron to remain in solution and be available to plants.

A highly alkaline soil having a pH above 8.0 is unfit for the growth of plants. The simplest and quickest way to make 100 ft² (9.2 m²) of this soil slightly acid requires the addition of 6–8 lb (2.7–3.6 kg) of ammonium sulfate.

To decrease the acidity or raise the pH of a soil, some form of lime is added. The three most commonly used are quicklime, hydrated lime, and ground limestone. Ground limestone is the safest of the three to use; when it is applied at a rate of 5 lb (2.2 kg) per 100 ft² (9.2 m²), the pH value will rise by three-fourths of a unit. This amount is equal to one ton (0.9 metric ton) per acre (0.404 hectare) of land. An additional benefit obtained from applications of lime is an increase of bacterial action within the soil.

In extremely acid soils, iron and aluminum may become toxic to some plants. Growth of fungi may increase under these same conditions at the expense of the soil bacteria, bringing their ability to decompose organic matter to a complete halt.

Soil *fertility tests* are usually conducted on soils to determine the available supply of the three most important chemical elements required by plants in order to live, grow, and reproduce. These three are nitrogen, phosphorus, and potassium. The amount in any soil of any of these plant food elements is extremely small and must be replaced periodically by fertilization.

In the past, most plant growers have assumed that their soil lacked all three elements and therefore applied a "complete fertilizer" containing all three elements. Presently, more growers are relying on a soil test with the realization that one or two of the necessary elements may be present in the soil in sufficient quantity and therefore may not be needed at all. The addition of a complete fertilizer would be both unnecessary and expensive.

Excessive amounts of fertilizer are extremely detrimental to all plants. The primary effect caused by this condition is the destruction of the root hairs. Once they are destroyed, the plant top begins to wilt and soon dies.

ADDITIONS TO THE SOIL

All soils benefit from the addition of humus or organic matter in a variety of ways. It increases the water-holding capacity of the soil, reduces the wide fluctuations in soil temperature, prevents the rapid leaching or loss of nutrients down through the growing area, promotes bacterial action, and improves the texture of most soils. Organic matter can be obtained from a variety of sources. They are

1. *Peat*. It is sold in a variety of forms. The two most common ones are the imported form, called peat moss, which is sold in bales by the cubic foot, and the domestic form, commonly referred to as peat humus. This type resembles soil and is generally sold by the pound.
2. *Compost*. This source of organic matter provides the home gardener with an excellent way of stockpiling and reusing all available plant wastes, such as leaves, weeds, and grass clippings. This method eliminates the need for garbage collection and at the same time provides a rich source of organic matter for the garden. (For more detail, see Chapter 27.)
3. *Manure*. It must be partially rotted before being applied to the soil. Once applied, it must be thoroughly mixed into the soil. Fresh manure, when used, should be placed on the soil surface for several months before being mixed into the soil. The fertility of manure is low, not high as some people mistakenly believe.
4. *Cover crops or green manures*. These include a variety of grasses

sown on bare ground for the prime purpose of increasing the organic content of the soil. Legumes are considered to be the best because of their nitrogen-fixing properties. All cover crops must be plowed under the soil before planting is started, making them ideal for most vegetable gardens.

5. *Leaf mold.* This form of organic matter makes an excellent soil conditioner. When obtained specifically from evergreens and oak trees, the leaf mold may be initially too acid but will soon be acceptable for all plants. If available, the material taken from the forest floor is the best to use.

6. *Wood products.* Sawdust, shavings, and wood chips can be placed directly into the soil, or they can be composted. Whenever they are used, fertilizer, especially one containing nitrogen and phosphorus, must be added because these wood products have low nutrient levels.

7. *Sewage sludge.* This is solid material obtained from the treatment of sewage and is sold in two forms, processed (activated) and raw (digested). Raw sludge should not be used on the vegetable garden unless it is left on the surface over the winter, since it may contain pathogenic organisms that can be readily ingested on all raw vegetables. Processed sludge has been heat treated and usually has no undesirable odor. It is frequently added to lawn surfaces for fertilizer purposes, but would be more effective if composted first.

SOIL PREPARATION PRIOR TO PLANTING

Before any soil can be planted, it must be properly prepared. There are certain required steps that must be carried out. They are

1. Tile fields and/or dry wells must be constructed to remove excessive surface water.

2. A minimum slope must be provided for efficient water removal.

3. Any utility lines and/or outdoor water systems that are to be permanently positioned should be placed below the existing frost line, which in New York state is about three feet.

4. Stones, rocks, and plant debris should be removed.

5. Drainage of the soil should be checked and corrected, if found to be poor.

6. All additions to the soil must be done before planting is started. They should be thoroughly mixed into the top 6–8 in. (15–20 cm) of soil.

7. If one is planting indoors, the soil must be pasteurized at a temperature of 200°F (94°C) for a minimum of 30 minutes. The soil must be moist when being pasteurized.

Chapter Three

Cultural Requirements of Plants

Before any plant can be grown successfully, the grower must have a complete understanding of its cultural requirements, which include

1. The intensity and duration of light needed.
2. The method of watering and its frequency.
3. The most nearly ideal growing temperature.
4. The frequency and method of fertilization.
5. The preferred soil type and most desirable pH.

This rule applies to the homeowner, the nurseryman, the florist, and the truck farmer alike.

The cultural requirements for most plants are merely suggested guidelines and thus provide the grower with sufficient flexibility to insure proper plant growth. However, there are certain plants that have very specific growth or cultural requirements which must be adhered to before success is possible. Under natural conditions, the successful grower must accept the environmental conditions that exist, but he must have the knowledge necessary to alter these conditions to guarantee ideal plant growth. For example, during periods of high light intensity, the grower should know when to shade and which plants must be shaded.

LIGHT REQUIREMENTS

All green plants produce some of their food by the process of photosynthesis. This unique process involves a chemical reaction in which plant cells manufacture carbohydrates from carbon dioxide and water in the presence of light and the green plant pigment, known as chlorophyll. The leaves are noted for most of the carbohydrate production, but stems and buds are also capable of photosynthesis. This process is totally dependent upon the plant's receiving quantities of light. Many plants thrive in complete sunlight, whereas others desire a partial or completely shaded environment. The grower must be aware of the best light environment for each plant grown and position the plants accordingly.

19

The plant symptoms associated with low light intensity are the following: leaf tips become discolored, leaves and buds drop, leaves and flowers become light in color, there is a lack of plant vigor, showing a decrease in growth, and the new growth appears stunted and weak.

The symptoms connected with too high light intensity are 1. the plant wilts, and 2. light-colored leaves may become gray in color.

Experimentation has shown that plants receiving continuous light of either high or low intensities become stunted and display a general lack of vigor when compared to plants given a normal light period followed by darkness.

Certain plants show a definite response to the length of their exposure to light. Knowledge of this response, known as *photoperiodism*, has enabled commercial growers to develop a program insuring production for a specific date or holiday. The poinsettia, popular Christmas potted plant, is known as a short day or long night plant, since it flowers naturally in December when the day length is short.

In New York state, the natural date for flower bud formation for the poinsettia is approximately October tenth. To ensure that the plants do not form a flower bud before this date and therefore flower much earlier than desired, the grower lights the plants during the middle of the night for four hours with small incandescent light bulbs. This action breaks up the long night into two shorter ones and keeps the plant in a vegetative, nonflowering state. Poinsettias grown in commercial establishments located on or near lighted highways must be covered with black cloth or plastic from October 10 until Thanksgiving to guarantee a saleable plant for Christmas.

WATERING REQUIREMENTS

Watering requirements vary with the type of plant, the texture of the soil, the time of year, and the growing temperature. During cloudy periods throughout the year, and during the winter months, light intensity is decreased, causing a decrease in growth. During these periods, plants require much less water. When transplanting any plant, whether it is a seedling or a mature tree, the addition of proper amounts of water helps to guarantee success.

The symptoms associated with underwatering are the following: crisp, brown leaf spots form, older leaves drop, leaf color becomes bleached, the plant wilts, flowers do not last and buds may drop, stems become weak and limp, suggesting a general lack of plant vigor, and the soil becomes dry and hard packed.

Overwatering may produce the following symptoms: leaves become curled and drop, leaf tips turn brown and die back, soft dark areas form on the leaf surface, new leaves grow soft and sometimes become discolored or even rotten, and the plant wilts because of root deterioration caused by a water-logged soil.

Some of the plant symptoms associated with the addition of too little or too much water are similar. In either case, the plant's top growth—leaves, buds, and stems—is deprived of the required quantity of water because of problems in the soil. Overwatering damages the root hairs by depleting their necessary oxygen supply, while too little water may totally dehydrate the roots, causing water to leave the roots rather than enter them because the concentration of water within the roots exceeds that in the soil.

The growing temperature is another important consideration. Plants grown outdoors must be able to withstand wide fluctuations in temperature and still survive, whereas indoor plants are usually more specific in their temperature requirements. When the growing temperature is specified for a plant, it usually means the night temperature that will produce the best plant. These same plants are usually not affected by daytime temperatures 10–15°F (3–4.5°C) above the specified night temperature.

TEMPERATURE REQUIREMENTS

Most plants, with the exception of cacti and certain succulents, which are characterized by large, fleshy leaves and stems capable of storing water, prefer a more humid atmosphere than the normal home provides. This condition can be corrected by positioning the plants near containers filled with water and away from drafts.

Symptoms associated with low humidity are injury to the leaf tip, wilting of the plant, and new growth that appears weak and distorted.

Plants grown under high humidity conditions may exhibit the following symptoms: soft dark areas on the leaves, plus soft, mushy, rotted leaf and stem tissue.

HUMIDITY REQUIREMENTS

Since all growing plants require fertilization, it is important for the grower to know when to fertilize (usually at the start of the growing or planting season), how much to give each plant, and what is the best method of application.

The amount of fertilizer that should be applied is dependent upon the natural fertility of the soil, the crop being grown, and the content of the organic matter present. For spring applications, the recommended amount for *lawns* is 1 to 2 pounds of nitrogen per 1000 square feet (10 to 20 pounds of 10-6-4); for *vegetable gardens* and *flower beds,* the fertilizer may be mixed with the soil prior to planting or afterwards spread between the rows or individual plants. *Shrubs* require 2 to 3 pounds of 5-10-5 for every 100 square feet, while *trees* should receive 1 pound of nitrogen for every 100 square feet of surface area. [It may be spread over the soil surface or injected into the soil at 18-inch intervals to a depth of 8 to 18 inches (.02 to .05 m), starting 2½ feet (.75 m) from the trunk.]

Plants differ greatly in the quantities of fertilizer that they require, but all are similar in that they must all receive the same chemical elements. From the list of twelve elements known to be essential for proper plant

FERTILIZATION REQUIREMENTS

growth, six are required in much larger amounts than the other six and are usually called the *macronutrients*. The three most noteworthy elements are nitrogen (N), phosphorus (P), and potassium (K). Each source of fertilizer has the soluble quantities of these three elements plainly printed on the package: i.e., a 5-10-5 fertilizer implies to the customer that the contents of the fertilizer bag has five parts nitrogen, ten parts phosphorus, and five parts potassium of the total weight. The other three macronutrients are calcium (Ca), magnesium (Mg), and sulfur (S).

The remaining six elements are needed by plants in very minute amounts and are, therefore, usually called *micronutrients*. They are iron, manganese, zinc, copper, boron, and molybdenum.

Fertilizer may be applied to the soil in granular form or in liquid form. In liquid form, it becomes much more readily available to the plant than in the granular form, which must be first broken down into a soluble form. The advantage to the granular form is that it is much longer lasting than the liquid form. Fertilizer may also be sprayed directly on the leaves of certain plants. Some fertilizers are labelled as *organic*, being slowly released to the plants, and others as *inorganic*, characterized by being water soluble and giving a fast response; others on the market are a combination of both types.

Overfertilization causes crisp dark spots on the leaves, lack of flowers on flowering plants, wilting of the plant, and root injury.

The symptoms caused by underfertilization are dropping of the older leaves, lack of flowers on flowering plants, pale bleached leaves, lack of plant vigor with stunted growth, and leaf veins yellow in color.

Table 3-1 lists the twelve essential chemical elements, giving specifics as to the function of each within the plant, and the deficiency symptoms associated with each.

SOIL pH

Receiving an ample supply of the essential chemical elements is no guarantee that a plant's growth will be normal. The degree of acidity or alkalinity of soils greatly affects growing plants. Most plants thrive best in the pH range of 6, which is slightly acid or sour, to pH 7, which is neutral, being neither acid nor alkaline. Survival may be possible for plants growing in an undesirable pH, but their growth is usually abnormal. For example, most commercial lawn grass mixtures are blended to insure that their preferences for soil pH are compatible. Kentucky bluegrass and the fescues are mixed together because they both prefer a pH near 7. Acid-loving grasses, such as bentgrass and red top, would suffer if grown in the same area where bluegrass is grown.

As mentioned earlier in the chapter on soils, a soil test kit can be used to determine the pH of a soil. It should be used chiefly as an indicator, suggesting to the tester what type of soil application is advisable. However, because of the complexity of most soils, the test may not be able to prescribe the actual quantities that must be added.

Table 3-1
IMPORTANT CHEMICAL ELEMENTS FOR PLANTS

Element	Function within the plant	Deficiency symptoms
Macronutrient		
Nitrogen	Involved in the formation of amino acids and proteins	Leaves light green, a decrease in height
Phosphorus	An important structural component of the nucleic acids (DNA and RNA) and fatty substances	Plant leaves dark green
Potassium	It activates several important plant enzymes	Leaf tips show dead areas, leaves mottled
Magnesium	A constituent of chlorophyll	Leaves light green and chlorotic, leaf margins turn upward
Sulfur	Found in certain plant amino acids	Areas between veins light green in young leaves
Calcium	Involved in cell wall formation	Plant tip will eventually die
Micronutrient		
Iron	A constituent of many important enzymes	Young leaves display chlorosis
Manganese	Involved in chlorophyll synthesis	Leaves are chlorotic with dead spots
Boron	Involved in sugar transport	Leaf tips become distorted, and eventually the tip dies
Zinc	An enzyme constituent	Dead spots on leaves that continue to enlarge in size, and leaves become thick
Copper	An enzyme constituent	Young leaves permanently wilted, tip of stalk unable to stand erect
Molybdenum	An enzyme constituent involved in nitrate reduction within the plant	Leaves light green

Additional cultural items that must not be overlooked by any grower are the spacing between plants, disease and insect control, cultivation of the surrounding soil, the presence of agents responsible for pollution, and the pruning, shaping, or shearing demands of each plant.

*Other Cultural
Requirements*

Many gardeners select their plants either for their colorful flowers or their edible fruits. In most instances, providing each plant with its proper cultural requirements is sufficient to guarantee the desired flowering and/or fruiting.

FLOWERING

Most plant researchers agree that the process of flowering involves the production of certain plant hormones or "auxins" within the plant. For certain plants, this hormone production is related to either a temperature or photoperiod response or to an interaction between the two, whereas other plants flower only after reaching a specific vegetative size. Some feel that flowering for these plants may be somehow related to their total leaf area, since they apparently are not dependent upon any specific growing conditions. However, once a specific plant has produced its own particular level of hormone, the process of flowering appears to be irreversible.

There are usually four stages associated with flowering:

1. The vegetative or nonflowering stage. It is characterized by active growth and stem elongation.

2. Flower bud initiation or formation. This stage is usually recognized by a cessation of vegetative growth and can only be totally confirmed by microscopic examination of the terminal bud.

3. Flower bud development. This is that segment of time between the actual flower bud formation and flowering. In most cases, this time interval is necessary for the enlargement and elongation of the bud.

4. Flowering.

These four stages must be thoroughly understood by the grower of any flowering plant to insure successful flowering. In the ensuing paragraphs, a more complete description as to how each of these four stages is attained with respect to those plants that respond to either temperature or photoperiod will be presented.

Temperature Response

The types of plants that flower because of their response to temperature are flowering trees, shrubs, bulbs, and fruit trees.

Stage 1. This occurs after flowering is over during the spring and early summer.

Stage 2. Flower buds form during the hot summer months.

Stage 3. Flower bud development occurs during the fall, winter, and early spring months during periods of low temperature above freezing and below 50°F (10°C). Knowledge of this phenomenon has made it possible for growers to force plants into flower before they do so naturally. This is readily observed at any indoor flowering show staged in the very early spring.

Stage 4. The natural time of flowering is approximately the same time each year.

Those classified as short-day or long-night flowering plants

Stage 1. To ensure that these plants remain in a vegetative condition, they are lighted with low wattage incandescent bulbs from 10:00 P.M. to 2:00 A.M. During late spring and throughout the summer the lighting is not necessary, because the days are long.

Stage 2. When the days are long, the plants must be covered with black cloth or plastic to achieve short-day conditions. Short-day conditions occur naturally during the fall months.

Stage 3. Once the flower buds are visible, the black covering can be removed.

Stage 4. It is now possible to have photoperiod-responding plants flower at any time during the year.

Long-day, short-night flowering plants

Stage 1. During the summer months, the plants are shaded to ensure the desired vegetative conditions.

Stage 2. During periods when the days are short, the plants must be lighted from 10:00 P.M. to 2:00 A.M.; they are grown naturally when the days are long.

Stage 3. The procedure is the same as for Stage 2 and is discontinued only when flower buds are visible.

Stage 4. Flowering occurs naturally during periods when the days are long.

Chapter Four

Tools

The selection of good tools is essential to everyone associated with the growth of plants. Whenever any tool is purchased, it must first be checked for durability, efficiency, construction, and ease of maintenance. Too often, a tool is chosen solely on the basis of a low price, with the rationalization that its performance should be fine for the small amount of use it will receive. If chosen with care, most tools should be long lasting.

CARE   Once chosen, tools must be properly maintained. They should be cleaned completely after each use. Boiled linseed oil should be applied to all wooden handles at least two to three times a year to preserve them. When left outdoors during the winter months and not treated, the handles become rough, occasionally split, and may present the gardener with splinters when the spring planting begins.

For sharpening the edges of any cutting tool, a file is a necessary item. Whenever large nicks are made in the cutting edge, a grinding wheel is usually faster and more effective than a file.

For the removal of dirt and rust from tools after each use and before winter storage, a wire brush is used. Once clean, all metal surfaces are treated with an oily cloth as a precaution against rust buildup over the winter months.

Pesticide sprayers and fertilizer spreaders should always be emptied and rinsed with water directly after their use. If chemicals are left in either, their longevity will be drastically reduced. A separate sprayer must be purchased whenever the broad-leaf weed killer 2,4-D or a related compound is to be used. It is imperative that the gardener label this sprayer accordingly and *never* use it for any other purpose.

Before any cutting tool is returned to storage, it should be cleaned, well oiled, and sharpened if necessary. By following this procedure, the tool is always ready for use, and unnecessary delays are avoided.

For the planting and transplanting of trees, shrubs, and gardens, a *long-handled shovel* [Fig. 4-1(a)] is more appropriate and versatile than a digging fork [Fig. 4-1(b)]. A *hand trowel* [Fig. 4-1(c)] is used for the planting of individual annuals, perennials, certain vegetables and bulbs.

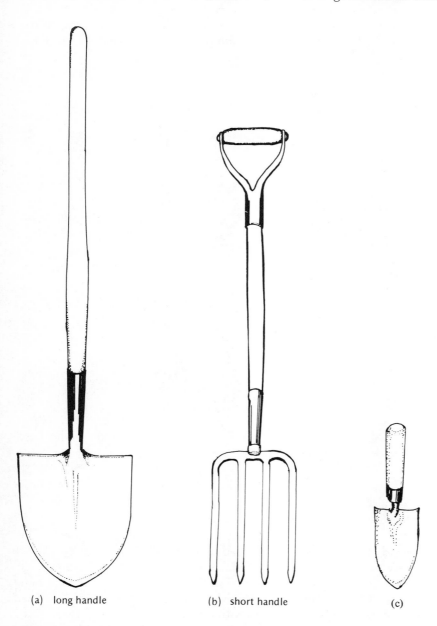

(a) long handle   (b) short handle   (c)

FIGURE 4-1
(a) Long-handled shovel. (b) Digging fork. (c) Hand trowel.

In grading and leveling a lawn or garden area, a *long-handled rake with straight tines* [Fig. 4-2(a)] should be selected. When held in an upright vertical position, it can be used very effectively to break up large dirt clods by drawing the tines back and forth across the clod. A *leaf rake* [Fig. 4-2(b)] is used to rake the fall leaves, to collect grass clippings, to work lime, fertilizer, and grass seed into a newly seeded lawn area, and to remove thatch, which consists of a buildup of grass particles on the lawn surface.

(a) long handle   (b) long handle

FIGURE 4-2
(a) Long-handled rake with straight tines. (b) Leaf rake.

A *hand pruner* [Fig. 4-3(a)] is used to prune small branches, to propagate woody plant material, and to reduce the stem lengths of both flowers and greens used in floral design work. For the pruning of large branches up to 1¼ in. (3.8 cm) in diameter, *lopping shears* [Fig. 4-3(b)] are needed. The shears selected should be designed with a stop just above the blades. The stop prevents the knuckles on one hand from colliding with those on the other hand. For hedges and simple pruning demands, *hedge shears* [Fig. 4-3(c)] are another necessary investment. They also should have a stop.

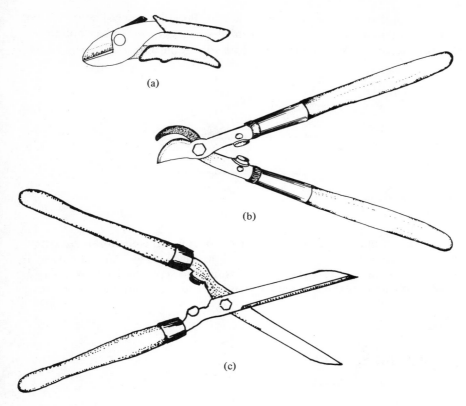

FIGURE 4-3
(a) Hand pruner. (b) Lopping shears. (c) Hedge shears.

For the removal of large tree limbs, a *pruning saw* [Fig. 4-4(a)] must be used. A straight saw with teeth on both edges is more versatile than the *bow saw* [Fig. 4-4(b)], whose effectiveness is limited to a definite tree limb diameter.

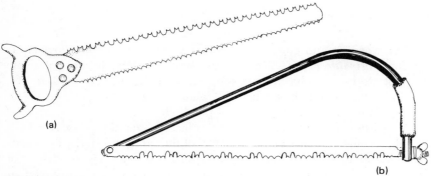

FIGURE 4-4
(a) Pruning saw. (b) Bow saw.

A tool that no gardener, propagator, or floral designer should ever be without is a *jackknife* (Fig. 4-5). Its small size gives every plant enthusiast the opportunity of keeping this tool with him at all times.

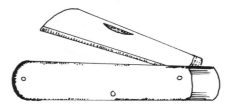

FIGURE 4-5
Jackknife.

Grass clippers [Fig. 4-6(a)] are used to trim and contain the growth of grass around permanent structures, such as shade trees, foundations, and sidewalks, while an *edger* [Fig. 4-6(b)] is used to separate one growing area from another, e.g., the foundation planting from the front lawn.

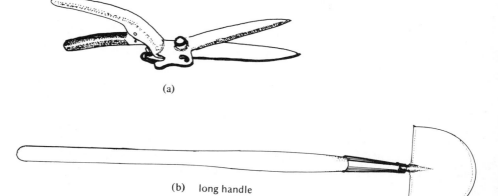

(a)

(b)   long handle

FIGURE 4-6
(a) Grass clippers. (b) Edger.

For weeding and cultivating around shrubs and among flowers, a *four tine cultivator* [Fig. 4-7(a)] is the best selection, because it does not penetrate too deeply into the soil root area. It can also be used effectively in the vegetable garden, but many prefer a *garden hoe* [Fig. 4-7(b)] to a potato hook at the time of planting.

No gardener should be without a sufficient quantity of *garden hose* that will adequately reach the outer limits of the property. In the formation of lawns and gardens or for simply watering a dry area, a variety of *sprinklers* are available, any one of which can be selected and positioned on the end of the hose.

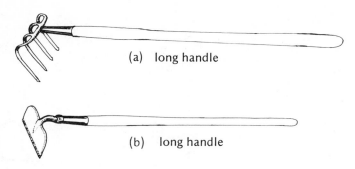

(a) long handle

(b) long handle

FIGURE 4-7
(a) Four tine cultivator. (b) Garden hoe.

A sprinkling can or *aerator nozzle* [Fig. 4-8(a)] attached to a hose is
used to apply a gentle, yet uniform quantity of water to propagation flats
containing either seeds or cuttings. Once the seedlings have reached the
proper size to be transplanted, a *dibble* [Fig. 4-8(b)] (a tapered wooden rod)
is used to make a hole in the container that will be the seedling's new
home. Some growers also use the dibble when sowing seeds.

*SPECIAL
TOOLS FOR
PROPAGATION*

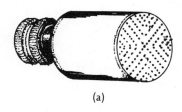

(a)

(b)

FIGURE 4-8
(a) Aerator nozzle. (b) Dibble.

The tools under this heading are not as frequently used as those listed
under basic tools. They are invaluable, however, whenever a problem
arises that requires their use to provide a solution.

*ADDITIONAL
GARDENING
TOOLS*

When planting must be done in hard or rocky terrain, a *pick-mattock*
[Fig. 4-9(a)] is selected. It can also be used for reforestation projects. The
large rocks that may be uncovered during planting are usually broken into
pieces with the use of a *sledge hammer* [Fig. 4-9(b)], before planting can
continue.

A *wheelbarrow* (Fig. 4-10) or garden cart is used to transport a variety
of garden items from one location to another. Some homeowners prefer the
wheelbarrow to the cart, because small quantities of cement or concrete
can be more easily mixed in the wheelbarrow.

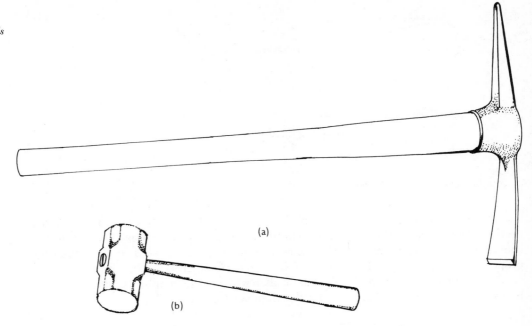

(a)

(b)

FIGURE 4-9
(a) Pick-mattock. (b) Sledge hammer.

FIGURE 4-10
Wheelbarrow.

For the simple maintenance of large trees, a *tree pruner* [Fig. 4-11(a)] capable of reaching 20-ft (6-m) limbs is used. It allows the gardener to trim hard to reach limbs while standing on the ground. A *root feeder* [Fig. 4-11(b)] may be used to fertilize large trees by supplying water-soluble nutrients at the root level. Some gardeners prefer to use a *crowbar* [Fig. 4-11(c)] instead of the feeder. They sink the bar 18–24 in. (45–60 cm) into the soil and move the bar back and forth to enlarge the hole. Once this is done, a handful of water-soluble fertilizer is poured down each hole, followed by sufficient water to prevent any damage to the roots. The crowbar can also be used in the removal of large rocks.

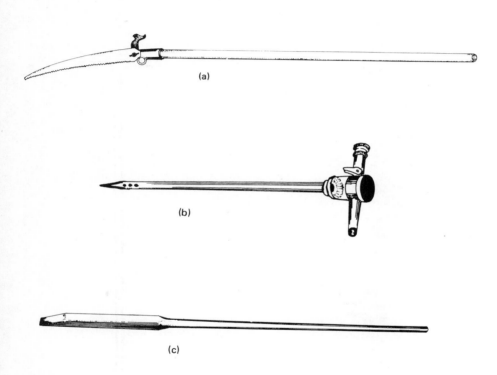

(a)

(b)

(c)

FIGURE 4-11
(a) Tree pruner. (b) Root feeder. (c) Crowbar.

Some gardeners prefer to use a *lawn sweeper* (Fig. 4-12) instead of a rake to pick up leaves and grass clippings on lawn surfaces. Most sweepers are readily adapted to be drawn behind a garden tractor so that mowing and cleanup essentially become one operation.

For fall bulb planting, a *bulb planter* can be used. It can be purchased in one of two sizes, as a hand tool [Fig. 4-13(a)] or one whose length is approximately the same as a short-handled shovel [Fig. 4-13(b)].

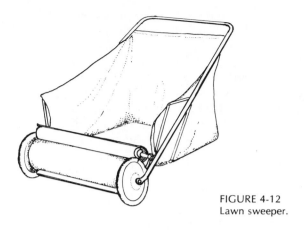

FIGURE 4-12
Lawn sweeper.

(a)

short handle

(b)

FIGURE 4-13
(a) and (b) Bulb planters.

A *post hole digger* (Fig. 4-14) can be used for digging holes for fence posts and the planting of certain trees and shrubs. When employed in rocky or hard-packed soil, the soil must be first broken up with a pick-mattock or crowbar.

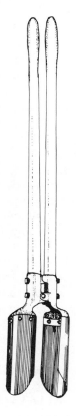

FIGURE 4-14
Post hole digger.

For concrete and cement work, a *mason hoe* (Fig. 4-15) and trowel are used. As previously mentioned, small quantities of either can be mixed in a wheelbarrow. The trowel selected may be either triangular or rectangular in shape and is sold in several sizes.

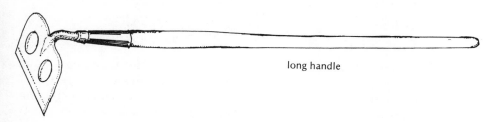

long handle

FIGURE 4-15
Mason hoe.

Chapter Five

Container Planting

The use of plant containers has greatly increased during the last few years. They provide the perfect solution to those apartment dwellers whose planting space is limited and to those homeowners who desire the flexibility of changing and rearranging their plantings to please their moods.

It is possible that the biggest problem that the purchaser may face with respect to container planting is to choose one container from a never-ending selection. Plant containers are sold in all sizes and shapes, and are constructed of a wide variety of different materials. Certain ones are designed specifically for propagation purposes, such as the germination of seeds or the rooting of cuttings, whereas others are constructed solely for the growth and display of plants. The quantity of containers selected for propagation is related to both the number of plants desired and the amount of available growing space.

CONTAINER CONSIDERATIONS Sometimes, when a specific plant has already been decided upon, the list of available containers is greatly reduced. Before the final selection is made, there are several considerations that must be taken into account:

1. The growth habit of the plant selected should be known.
2. The container's overall dimensions should be large enough to provide for proper root growth.
3. The container should have some provision for drainage.
4. The relative ease with which the container and plant can be moved must be considered.
5. The material used in the construction of the container should complement the container's proposed location.

With the increasing use of masonry materials for the newer contemporary homes, more permanent types of plant containers have been constructed as an integral part of the landscape. These include: (1) built-in planting beds and (2) plant pockets or holes large enough to contain plants left in the patio surface.

One of the strong points of container planting is that a soil mixture can be prepared that will satisfy the specific requirements of each plant grown.

All container plants should be watered with care and with the realization that overwatering is as detrimental as not enough water. Containers that have no provision for the drainage of excess water must be initially planted properly to avoid damage to plant roots. For example, containers such as terrariums, dish gardens, and undrained glazed pottery all require a layer of gravel in the bottom of the container. This gravel layer must be at least one in. (2.5 cm) thick. On top of the gravel should be placed an inch (2.5-cm) layer of sand or a thin layer of sphagnum moss before soil is added to the container. This procedure, coupled with the proper watering technique, should keep the plant material in these containers alive and healthy.

Fertilization is best accomplished by adding a complete fertilizer either in liquid or in granular form. (See Chapter 4.) During periods of lower light intensity or decreased plant growth, the frequency of fertilization should be diminished.

In this chapter, containers will be divided into two categories: those that can be moved from one place to another, and those that are permanent.

A glass jar positioned on its side (Fig. 5-1) is excellent for the germination of very fine seed, such as members of the gesneriad family (Chapter 38). The air holes placed in the jar's top must be above the soil line. This is determined before the soil is added by placing a mark on the glass where the cover stops when it is tightly closed. Since most jar tops are designed to return to the same position when closed, the mark will help determine the soil's position within the jar. Once this is accomplished, the holes are made and restricted to the upper half of the jar. The soil is screened, placed in the lower half of the jar, and moistened with a mister or atomizer. The seed is sown by first placing it on a spatula and simply turning the spatula over inside the jar. The top is then placed on the jar.

A plugged 2¼-in. (5.7-cm) clay pot placed in the middle of a larger 6–7-in. (15-18-cm) pot is very effective for seed germination, since it provides the grower with an automatic watering system. The larger pot, once

MOVABLE
CONTAINERS

FIGURE 5-1
Glass jar on its side.

filled with soil and properly moistened with water, is sown with the desired seeds. After the seeding has been completed, the smaller pot is filled with water. Because of the porosity of the clay, the water flows freely through the sides of the smaller clay pot into the surrounding soil area, ensuring that the germinated seedlings in the larger pot never dry out. To make the container more efficient, it can be enclosed in plastic (Fig. 5-2).

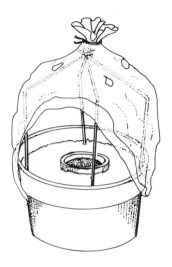

FIGURE 5-2
Pot within a pot.

A variety of containers used for propagation purposes are constructed of a peat moss composition; these are designed to house the propagated plant material until it is planted and then to disintegrate. Jiffy pots are individual round 2¼- or 3-in. (5.7- or 7.6-cm) pots. Jiffy strips are 12 attached square pots that are easily separated at planting time. Jiffy flats are available in two popular sizes, 5½ × 7½ × 2¼ in. (14 × 19 × 5.7 cm) and 10 × 12 × 2½ in. (25 × 30 × 6.4 cm) and can be used to house a variety of plants until planting time (Fig. 5-3).

FIGURE 5-3
Jiffy flat.

Jiffy 7's are seed starter pellets that expand to seven times their original height when placed in water. They are 2 in. (5 cm) in diameter and are composed of peat moss, a sterile planting mix, and fertilizer, and are designed to provide the germinated seedlings with all their needs until planting time (Fig. 5-4). Jiffy 9's resemble jiffy 7's, but are only one inch (2.5 cm) in diameter.

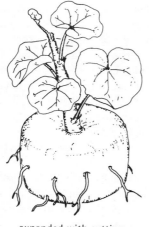
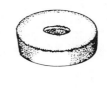

expanded with cutting

FIGURE 5-4
Jiffy 7's.

Seed starter cubes have a wood fiber composition and are available in three hole sizes for different-sized plants, such as seeds or stem cuttings (Fig. 5-5). Because these cubes are designed to stay constantly moist, they frequently shorten the normal germination time required for most seeds. They are also designed to supply the newly germinated seedlings or rooted cuttings with fertilizer over an extended period of time. When planting time arrives, the cube and its contents are planted intact. This procedure enables the plant's continued growth without the setback associated with transplanting.

FIGURE 5-5
Seed starter cubes.

Clay pots are available in a wide range of sizes and depths. The standard size pot has the same depth and diameter, while azalea pots are three-fourths the depth of a standard-sized pot. Bulb pans are only half the depth of a standard pot. Clay pots are extremely porous and, when new, must be thoroughly soaked before being used. Because of this porosity, plant roots congregate just inside the pot. These pots have only one drainage hole in the bottom.

Plastic pots are sold in the same sizes as clay pots, but are much lighter in weight. They are not porous, and therefore plant roots form throughout the entire soil mass. Most plastic pots have four drainage holes.

Flats may be constructed from either wood or plastic. When used for propagation purposes and for growing plants, they must have some provision for drainage. The standard wood size is 16 × 22 in. (41 × 56 cm); the normal greenhouse bench size is 12 × 18 in. (30 × 46 cm); the shelf size is 6 × 9 in. (15 × 23 cm). For the germination of seeds or the rooting of cuttings, the flat should be two to three inches (5–7.6 cm) deep. When one is growing cut flowers or forcing bulbs, the flat has to be at least 6 in. (15 cm) deep.

Watertight flats are used to house potted plants that will be left unattended for two to three days (a weekend) by supplying the necessary water from below. They are also used to raise the humidity level required by specific plants on a more permanent basis. These plants are placed on stones within the water-filled flat, thus insuring against possible damage to the roots, which could result if the plants were permanently left in water.

Hanging baskets have become extremely popular as a means of displaying plants. They may be constructed out of clay, plastic, or ceramic. Some types, such as the ceramic pots, have no provision for drainage and need an inch (2.5 cm) of gravel, topped by another inch (2.5 cm) of sand before planting can occur. Some hanging baskets are constructed from wire and must be lined completely with coarse sphagnum moss before receiving the plant and its soil mixture. This type must be carefully watered so that the exterior moss cover remains intact.

Another popular method of displaying plants is the terrarium or glass garden. These glass or plastic containers come in a variety of sizes and shapes. To insure drainage, one inch (2.5 cm) of gravel must be placed in the bottom of the container, followed by an inch (2.5-cm) layer of sand. To keep the soil from becoming too acidic, granulated or powdered charcoal, at the rate of one tablespoon per 4-inch (10-cm) pot of soil, is mixed with the soil. Most terrariums are designed with a cover, which, when kept in place, should guarantee a perfect water cycle. However, sometimes condensation appears on the inside of the glass, requiring removal of the cover to allow the excess water to evaporate. Proper plant selection is important for terrariums, since not all plants are suitable. (See Chapter 32 for additional information.)

A shallow dish, first lined with gravel and then filled with soil, is known as a dish garden. Because the depth of the dish is slight, the selection of plants is restricted, usually to succulents and cacti. Frequently the soil surface is covered with sphagnum moss or small colored pebbles for a better appearance.

Bonsai is a Japanese art designed to keep a plant small by pruning and wiring its branches (see Chapter 36). The containers used for bonsai resemble dish gardens in depth, but all have holes for drainage.

Rectangular planters that have no provision for drainage require the same treatment mentioned above for hanging baskets. The length of these containers may vary, but the depth and width are usually the same. When the depth is only four inches (10 cm) and the drainage materials take up half that depth, the plants selected must either be small or have their root systems decreased to fit the depth.

Portable boxes differ from planters in that the depth, length, and width are all very similar in their dimensions, but are planted in the same manner.

Plant tubs and urns can be purchased in many sizes and shapes and may be constructed from redwood, concrete, or pottery. Some have provision for drainage; those that do not must be planted accordingly.

Built-in planting beds are constructed either to blend with or to be of the same material as the house. Provision for drainage of excess water should be made at the time of construction, and the house must be protected from excessive moisture from the bed by employing flashing between the house and the bed. *PERMANENT CONTAINERS*

Plant pockets are spaces left in the patio surface to hold specific plant material. Before the patio surface is positioned, drainage for these pockets must be completed.

Outdoor window and porch boxes are best constructed from fiberglass, cypress or redwood. The best dimensions for these boxes are: length, 32–48 in. (80–120 cm); width, 10 in. (25 cm); depth, 8 in. (20 cm). They are planted in the same manner as the rectangular planters. All three materials are strong and durable. The weight of the soil in any window box attached to the exterior of the house is sufficient to necessitate a firm support for the box. In most instances, the easiest solution is to position at least three support brackets under the box, one at each end and the third in the middle. For those individuals who object to the use of brackets, a minimum of six lag screws can be used. They must be at least 3 in. (7.5 cm) long, and must be positioned in pairs, one near the top of the planter and the other near its base, to catch three consecutive structural two by fours (2 × 4's). If this task is to be performed properly, a wrench must be used to anchor the lag screws to the 2 × 4's.

Greenhouse benches may be constructed of cypress, concrete, flakeboard, pressed asbestos, or fiberglass. The material decided upon involves a balance among cost, durability, and the type of plants to be grown.

Chapter Six
Watering Plants

Most plants have the ability to adapt to environmental conditions that are less than ideal, but without water they soon wither and die. In certain locations, plants receive sufficient water as rain, but in most places and at certain times of the year, the plants must receive supplemental amounts of water.

The botanical reasons for watering are the following:

REASONS FOR
WATERING

1. *To prevent wilting.* Without sufficient quantities of water throughout the entire plant and especially in the cells of the leaves, photosynthesis is curtailed and the leaves become limp and begin to droop.

2. *To reduce evaporation at the leaf.* The evaporation of water from all above-ground parts of the plant is called *transpiration*. It is greatest in the leaves during periods of low humidity and turbulent air. It is slowed greatly when the stomata are closed. The stomata open with illumination and a decrease in CO_2 concentration near the leaf. Figure 6-1 shows the cross section of a leaf and the presence of a stomate, surrounded by two guard cells.

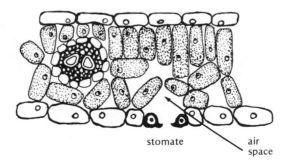

stomate   air
space

FIGURE 6-1
Cross section of a leaf.

3. *To dissolve the salts present in the soil.* The nutrients required by plants for all their vital functions can be absorbed by the root hairs only when these nutrients exist in a soluble form. Without water, they would remain in the soil in an insoluble form.

4. *To insure the movement of plant fluids.* Nutrients and materials within the plant can be mobile, moving from one cell to another, only when there is sufficient water in the plant tissues. The flow of these materials ceases whenever water is lacking.

5. *To maintain the plant's turgor pressure.* Within each plant cell water exerts pressure against the cell walls, making certain that the cells maintain their normal rigid or bulging condition. Without this pressure, cell activity decreases and eventually stops completely.

6. *To maintain a humid atmosphere.* As the humidity levels in and around a plant increase, the rate of evaporation or transpiration from the leaf surface is remarkably decreased, and consequently the plant requires less water.

Most plants need about an inch of water a week during their growing season and even more when flowering and fruiting occur. To supply plants with the equivalent of one inch (2.5 cm) of rainfall during dry periods, each square foot (960 cm²) of soil should receive two-thirds of a gallon (2½ liters) of water. It is advisable to supply this quantity slowly over a long period of time. When applied slowly, the water is able to penetrate deeper and more completely than when applied rapidly.

The water requirements for most plants vary with:

1. *The stage of growth.* More water is needed by plants at flowering and fruiting times than at all other times. Plants in a dormant state seldom need water, and care must be exercised during this time. Too much, however, applied during active growth may result in the cracking of fruits and possibly even plant death.

2. *Environmental changes.* As the air temperature and light intensity increase, causing most plant reactions to also increase, the plant's demands for water are greater. On cool, overcast days, most plants may not require water.

3. *Type of root system.* Experimentation has proven that plants having fibrous-type root systems require more water than those having tap roots. Most greenhouse bench plants form a vigorous root system extending deep into the bench when they receive heavy, but infrequent waterings.

4. *Texture of the soil.* Clay soils are able to retain water longer than sandy soils. The average loam holds about 2½ in. (6.4 cm) of water per foot (31 cm).

For some gardeners, knowing why plants need water is not nearly as important as knowing when to water their plants. There are now commercial *soil moisture meters* available that take the guesswork out of watering.

A *rain gauge* placed in the garden away from large trees is another instrument that will tell the gardener exactly how much moisture the plants have received.

For plants grown in clay pots, an old reliable method used is to tap the side of the pot; if a hollow ringing sound results, the pot should receive water.

When plants growing in a greenhouse bench or lawns and gardens outdoors are watered, the soil should be checked at a depth of ½ inch (1.25 cm). If the soil is light in color, the area should be watered.

Problems
Associated
with Watering

Fungal disorders, such as crown rot and leaf spot, are best prevented by providing adequate ventilation and attempting to keep the plant dry. The problem intensifies whenever the air temperature and the humidity level rise.

Root damage usually results from overwatering, which restricts the available oxygen in the soil. Once this occurs, the plant suffocates from lack of oxygen and usually dies.

METHODS OF
WATERING
PLANTS
A. In the Home

For simplicity, this section will be presented in outline form, with explanations and diagrams being given where needed.

1. *Water faucet.* For best results, the water should be at room temperature and applied just inside the rim of the pot at low pressure.
2. *Sprinkling can or mister.* Both are excellent for new seedlings and for raising the humidity level. Neither should be used on plants whose leaves are prone to fungal disorders.
3. *Submerging the entire container.* Potted plants are placed in a container of water and thoroughly saturated with water. They are lifted from the container and the excess water is allowed to run off.
4. *Watering from below.* Pots are placed in shallow watertight containers and removed as soon as their soil surface feels damp.
5. *Pans of water filled with stones.* Since the pots rest on the stones and not actually in the water, this method does more to increase the humidity level than it does to water plants.
6. *Wicks.* This is an automatic method of watering clay pots. One end of the wick (cotton clothes line will work) is pushed through the drain hole in the pot and flared out over the bottom of the pot. The other end is placed in a small reservoir of water.
7. *Terrarium effect.* This method involves enclosing the entire plant in a plastic bag. It is ideal for those plants left unattended for vacation periods up to one week long. This method is also very effective in raising the humidity level.

B. In the
Garden and
Lawn

1. *Garden hose.* It may be used manually or set to run automatically. In either case, it is advisable to apply the water slowly rather than rapidly.

2. *Oscillating sprinklers*. These are commercially available in a variety of designs, and they can be set up to run automatically.

3. *Soaker hoses*. These may be placed on the surface or below the surface; they are very effective at supplying small quantities of water over long periods of time. If left in place for the winter, they must be drained completely.

4. *Root waterers*. These are connected to any garden hose and are excellent at providing the roots of large trees with water. A minimum of eight holes should be used for each large tree. Each hole should be at the outer branch limits of the tree and should be at an angle of 45 degrees with the next one.

5. *Permanent watering systems*. These are usually put in position when the house is being constructed. They must be placed below the frost line and must be completely drained for the winter.

*C. In the
Greenhouse
(automatic devices)*

1. *Mist*. This method is used primarily for propagation purposes. With the system shown in Fig. 6-2, a stainless steel screen is used to turn on and shut off the system. Once the screen is saturated with water, it becomes heavy and drops, shutting the system off. As the water evaporates on the screen, it rises and causes the misting cycle to begin again.

2. *Tube weights*. These are also known as drip or trickle irrigation. Small-bore polyethylene tubing carries the water to a tube weight. The desirability of this system is that the water can be applied steadily or intermittently, a drop or trickle at a time. It finds its greatest use among pot plant growers.

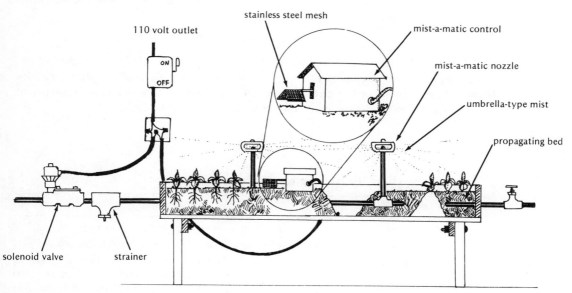

FIGURE 6-2
Automatic mist setup.

3. *Soaker hose, twin wall hose, and ooze hose.* All of these are placed on the greenhouse bench at planting time and are used mainly for the growth of cut flowers.

4. *Hydroponics.* This method involves the soil-less culture of plants grown in nutrient solution. The dangers associated with this method are twofold. The spread of disease may totally destroy the entire crop, and the nutrient concentration may either become too great or too low, requiring constant checking.

5. *Syringing.* This is a *manual* method that involves the use of a high-pressure force of water. It increases the humidity, thus reducing transpiration. It also helps in the control of red spider mite infestations.

LIST OF DON'TS

Do Not

1. Use very cold water.

2. Apply water to the tops of plants with hairy, fuzzy leaves, such as African violets. These should be watered from below or with extreme care.

3. Allow plants to stand in containers of water for long periods of time.

4. Water outdoor plants as the temperature is rapidly dropping.

5. Grow any plant in a container that has *no* provision for drainage.

6. Lightly water any newly germinated seedlings, i.e., lawns, vegetables, or annuals. The growing area must be well saturated with water when needed or left entirely alone to receive only water in the form of rain.

AVAILABLE WATER SOURCES

Water from a variety of sources is used to satisfy the watering requirements of plants. Each water source usually contains certain impurities that may affect the optimum growth of plants. Table 6-1 gives a listing of the various water sources and comments related to each.

Table 6-1
SOURCES OF WATER FOR PLANT GROWTH

Source	Comments
Rain water	The best water source for all types of plants.
Distilled water	This form is laboratory pure but too expensive.
Chlorinated water (urban or city)	When this form is used for house plants, it should be placed in a nonmetal container for 12 to 24 hours before use. Plants may be damaged if the free chlorine content becomes excessively high.

Table 6-1 (continued)
SOURCES OF WATER FOR PLANT GROWTH

Source	Comments
Softened water (de-ionized)	This source may cause plant damage if the sodium content is too high. It is also expensive.
Laundry water	It can be used to supplement the normal water source without ill effects.
Irrigation water	This type should be tested periodically to make certain that the soluble salt content remains low.
Untreated (pond, well, river) water	It is advisable to have any untreated water source tested before applying it to plants.
Water with high levels of copper, lead, and/or zinc	This source may be toxic to certain plants.
Hard water	Its use may cause a buildup of calcium (lime) and magnesium ions in the soil, resulting in an increase in pH to a level too high for suitable plant growth.

Chapter Seven

Methods of Propagation

SEED HYBRIDIZING   A variety of propagation methods are employed to obtain new plants. The only sexual method used involves the fertilization of the ovary or female part of the flower by the pollen from male anthers (Fig. 7-1). Most hybrid seeds are obtained by selecting the pollen from a plant with desirable characteristics and transferring it to the ovary of another plant with other desirable characteristics. The flower head is then completely covered with a paper bag to ensure that the ovary can be fertilized only by the pollen selected and can not be cross-pollinated by wind-borne pollen. The anthers on the flower head selected are removed before the paper bag is put in place and before they shed their pollen.

Propagation from seeds

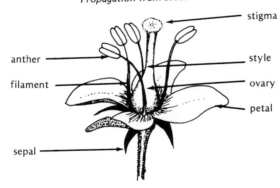

Male: stamen, consisting of
    anther (contains pollen)
    filament (for support)

Female: pistil, consisting of
    stigma (traps the pollen)
    style (tube to the ovary)
    ovary (contains the egg(s))

FIGURE 7-1
Typical flower diagram.

It is the intention of the hybridizer to produce a seed that has the desirable traits of both parents.

All other methods of propagation are classified as *asexual or vegetative*, whereby some portion of the established plant is used to obtain a new plant, which will be identical to its parent in all respects.

The seed packets sold to the public are the culmination of years of cross-pollination to obtain the best possible flower and vegetable plants. These packets are inexpensive and abundant, and they contain seed that is disease free and has a high percentage of germination.

For the purchaser to be successful with the sowing and germination of seeds, the following suggestions and requirements must be met:

1. Use a sterile medium and a sterile container.
2. Purchase healthy seeds from a reliable source. Seeds do stay viable for several years if stored at a temperature below 60°F (15°C) and at a humidity level of about 20 percent, but the average gardener has difficulty maintaining these conditions.
3. Completely fill the container with the medium. This practice insures better air circulation, thus restricting the growth of fungal organisms.
4. Thoroughly moisten the medium before sowing the seeds. Water flowing out of the drainage holes is a good indication that this has been accomplished.
5. Use a plant label to mark out the rows, remembering to give ample space between each row.
6. Plant most seeds at a depth three times the diameter of the seed. Be careful not to cover any seed too deeply. Very fine seed should be pressed very lightly into the medium.
7. Provide larger seeds with more space between each seed than you do for smaller seeds.
8. Once the seed flat is sown, cover it with either a pane of glass or moist newspaper to conserve moisture. Those seeds that require light for germination will have to be watched more carefully to make certain that they do not dry out.
9. Enclose the entire flat in plastic whenever it is anticipated that the loss of moisture may become excessive. Be sure to use stakes or plant labels to keep the plastic off the medium.
10. Make certain that the seed flat is not subjected to poor air circulation or too much moisture, since these conditions may encourage the growth of soil mold.
11. As soon as germination is apparent, move the flat into full light and remove any covering used.
12. Maintain temperature range of 65–70°F (20–21°C) for the germination of most seeds.
13. Water new seedlings with care by either soaking the complete flat in a pan of water or by using an extremely gentle sprinkling from above. Make certain that the new seedlings never dry out.
14. Realize that watering requirements will vary from day to day for new seedlings. On bright sunny days, they may need to be watered more than once, whereas on cool, cloudy, overcast days, they may not require any water.

15. Transplant seedlings to a more permanent location once they have developed one to two sets of true leaves.

16. Make certain to lift seedlings from below so that their small roots are not injured. This is best accomplished by using a plant label or your fingers.

Many house and garden plants are easily obtained from the parent plant by propagation from stem cuttings. The procedures used to obtain the three major types are as follows:

I. Softwood
Herbaceous
Plants (Fig. 7-2)

Begonias, chrysanthemums, coleus, geraniums, and a variety of foliage plants are obtained from stem cuttings.

1. Select plant tips with a minimum length of 3 in. (7.5 cm).

2. Remove the cutting from the parent plant by either cutting it with a knife or breaking it with your fingers, making certain to make the separation between two adjacent leaf axils.

3. Remove those leaves that will be placed below the surface of the rooting medium in order to minimize any loss due to the presence of disease organisms.

4. If faster rooting time is desired, dip the bottom of the stem in a rooting hormone.

5. Once the cuttings are placed in the rooting medium, keep them moist but not drenched.

6. Plant the cuttings as soon as they offer resistance when given a slight upward tug.

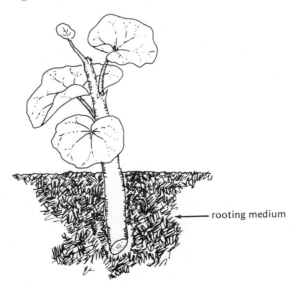

rooting medium

FIGURE 7-2
Herbaceous stem cutting—geranium.

Forsythia, privet, and dogwood are examples of this type.

1. Obtain the cuttings during the fall from the current season's growth.
2. Select cuttings 6–10 in. (15–25 cm) long with three to four buds per stem.
3. Either cut the base of the cutting on a diagonal, or make sure that it contains a portion of more mature wood; i.e., make either a heel or a mallet cutting.
4. Tie the cuttings in bundles, label them, and bury them in moist sand until spring.
5. When the cuttings have become sufficiently calloused—i.e., swollen at the base—plant them in rows outdoors.

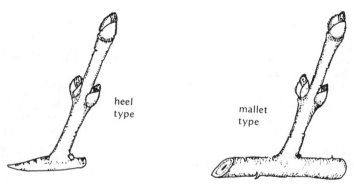

heel type

mallet type

FIGURE 7-3
Hardwood dormant cuttings—heel and mallet types.

Japanese yew and a variety of junipers are propagated in this matter.

1. Take the cuttings during the fall months from the current season's growth.
2. Place the cuttings in a flat in the greenhouse at a temperature of 60°F (13°C).
3. Hasten rooting by keeping the humidity high and protecting the cuttings from direct sunlight.
4. Once rooting has taken place, transplant the cuttings into pots or plant them directly in the nursery in the spring.

Many house plants are propagated from *leaf cuttings*. The general rule of thumb that is used by most propagators is: the thicker the leaf, the easier and more successful will be the propagation. The three methods used are the following:

A. Petiole leaf cutting (Fig. 7-4). African violets and gloxinias are propagated by this method, which requires that the leaf petiole or stem of the

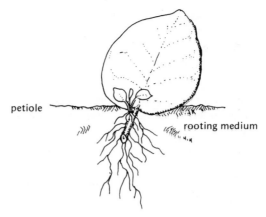

petiole

rooting medium

FIGURE 7-4
Petiole leaf cutting—African violet.

leaf be immersed in the medium until rooting occurs. The leaf itself rarely touches the rooting medium.

B. Whole leaf (Fig. 7-5). This method of propagation is used to start new plants of rex begonia. The entire leaf is placed flat against the rooting medium and is weighted down with small pebbles. When the leaf is properly positioned, its main veins are cut completely through. New plants will form where these cuts were made.

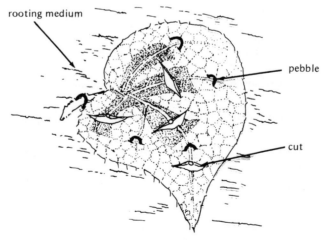

rooting medium

pebble

cut

FIGURE 7-5
Leaf vein cutting

C. Partial leaf (Fig. 7-6). This is another method of obtaining new rex begonia plants. It involves making V-shaped sections of the entire leaf and placing the tip of the section in the rooting medium. Each section must

FIGURE 7-6
Leaf triangle cutting

contain a large vein. Snake plant can be propagated by placing the tip of a one-inch section in the medium.

Besides the house plants already mentioned, the following are also propagated from leaves: aloe, croton, kalanchoe, peperomia, pilea, sedum, and wax plant.

Dieffenbachia, dracena, and rubber plant are three popular house plants propagated by *air layering* (Fig. 7-7). This same procedure can be used to propagate new plants from hardy flowering trees and shrubs, such as dogwood, rhododendron, and holly. The technique involves the use of

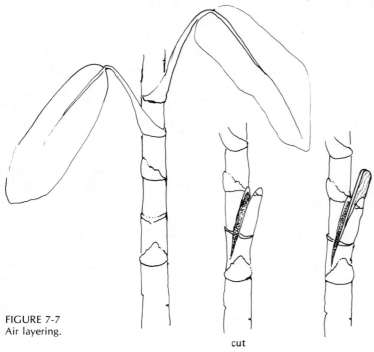

FIGURE 7-7
Air layering.

cut

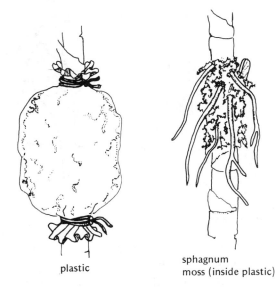

FIGURE 7-7 (Continued.)   plastic   sphagnum
                                                 moss (inside plastic)

sphagnum moss enclosed in plastic to induce new roots to form in the area where the stem has been partially cut. For success, the following steps must be followed:

1. Thoroughly moisten the sphagnum moss for two days before starting the air layering.
2. Cut a notch or ring the stem in the area to be layered. Place a toothpick or small wood chip in the notch to hold it open.
3. Pack the area, both above and below the cut, with the sphagnum moss. Make certain to squeeze the excess water out of the moss.
4. Cover the area selected with clear plastic and seal the plastic and moss completely with black electrical tape.
5. Once roots are visible through the plastic, sever the new plant from its parent and plant it.

Plants with branches nearly touching the ground are easily propagated by *ground layering* (Fig. 7-8). Forsythia and black raspberry are obtained by following the procedure below:

1. Select a stem that is growing close to the ground and cut a notch in it.
2. Depress the notched stem below the soil surface and cover it completely with soil. It is advisable to use a small forked stick to keep the stem in place.
3. Cut the new plant away from its parent after one growing season and plant it in a more permanent location.

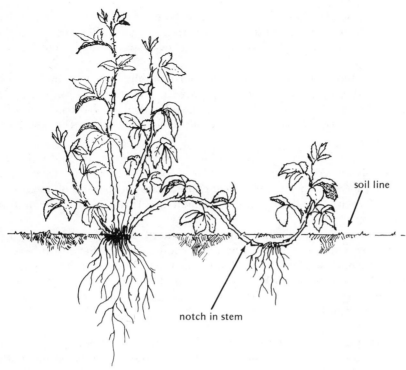

soil line

notch in stem

FIGURE 7-8
Ground layering.

Serpentine layering, a compound type of ground layering, is used to propagate certain vines having long flexible stems capable of sending up shoots from those buds on the buried part of the stem. Ivy and willow may be propagated by this method which yields three to four new plants per stem.

Many herbaceous perennials and a few shrubs may be propagated by *division of their clumps.* The clump must be dug up and separated with a garden spade or sharp knife and replanted, the new plants being kept well watered for the first few weeks. With the exception of chrysanthemums and asters, most perennials should be divided in early autumn after flowering is completely finished.

Certain perennials are propagated by *root cuttings.* The plant must be dug up before the 2–3-in. (5–7.5-cm) root section can be obtained. Most root cuttings are obtained during the fall, planted in flats of sand, and placed either in the greenhouse or in a cold frame for the winter. They should be ready for planting in the spring. Oriental poppy, phlox, and baby's breath are usually propagated by this method.

Bulbs are composed of fleshy, scalelike leaves and modified stems, both found growing beneath the ground. They contain large quantities of stored plant food. There are two general types of bulbs: those whose leaf tissue layers are compact and covered with a dry husk, like the tulip, narcissus, and hyacinth, and those with thick, overlapping scales, characteristic of lilies.

Many bulbs are increased in number naturally from a mature mother bulb. It is dug up as soon as the top growth shows signs of becoming dry and discolored. The new bulbs are removed and planted directly into the ground or into flats. It usually takes about three years before the new bulbs will flower.

A *corm* is a short, thick, round underground stem. Its fleshy tissue is solid and is *not* composed of leaves. New corms of flowering size are produced from the mother corm, which disintegrates at the end of its growing season. Crocus, gladiolus, and cyclamen are three plants grown from corms.

A *tuber* is a thick, enlarged stem section that grows underground. It has small scalelike leaves and tiny buds known as *eyes*. Several tubers are produced from each section containing at least three eyes. The eyes give rise to the new plants, which obtain their food while growing from the tuber section. Potatoes, dahlias, tuberous begonias, and artichokes are all grown from tubers.

PROPAGATION BY
GRAFTING
(Fig. 7-9)

Grafting involves the insertion of a shoot (scion) from one plant into the slit (stock) of another plant so that the scion will grow permanently. Grafting is done to:

1. Save damaged trees.
2. Increase production yields.
3. Insure these new plants a hardier root system.
4. Perpetuate hybrid varieties whose seed would be different from the parent plant.

Budding or bud grafting (Fig. 7-10) is the fastest method available to increase the number of a desired variety. One stem from a newly developed hybrid tea rose containing 15 buds can yield 15 new rose plants.

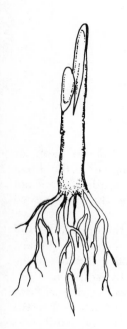

(a) stock (old)

(b) scion (new)

(c) union of stock and scion

FIGURE 7-9
Grafting—stock and scion.

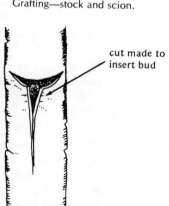

cut made to
insert bud

bud

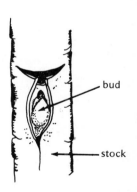

bud

stock

stem with buds

FIGURE 7-10
Bud grafting.

Chapter Eight

Seed Production and the Treatment of Seeds

Whenever a home gardener purchases a package of flower, lawn, or vegetable seeds, he is confident that the seeds will germinate if the directions on the seed packet are followed. However, most gardeners are not familiar with all the requirements that must be met by the various seed producing companies prior to the final packaging.

Most flower and vegetable seeds are produced in the inner valleys of California, which provide favorable growing and harvesting conditions, consisting of a mild climate with limited amounts of rainfall. Because of the lack of rainfall, most seed-producing fields are watered by irrigation.

Plant breeders are constantly cross-pollinating a wide variety of flowers in test fields in an attempt to provide the home gardener with seeds having improved characteristics, which can range from larger-size flowers or fruits to total resistance to a specific disease organism. They must keep extensive records of the various crosses they make, so that whenever they produce a new, more desirable variety, they are able to duplicate their results for the mass market. The entire process of developing new varieties is both complex and time-consuming and requires critical examination by the plant breeder during the new seed's first year of growth.

Only about 5 percent of the total number of crosses made each year are chosen for further testing and study, and it may require 12 to 15 years of continuous cross-pollinating to achieve the most desirable plant.

Seeds that have finally reached the stage of public acceptance are planted continuously from November to May. Each seed grown, however, has its own specific schedule of planting, growing conditions, pollination, and harvest. The growth requirements or culture of one species of seed may be entirely different from another species; therefore, the knowledge of a well-trained grower is required to produce the best seed possible.

The grower must also know whether the seed species entrusted to him are to be self-pollinated by either wind or insects or cross-pollinated by hand, which requires the emasculation or removal of the male components of the flower and the transfer of the desired pollen from other plants grown in another location to the ripened pistil. Seeds obtained in this manner are

58

given the designation of hybrid seeds (F$_1$ hybrids) and must be harvested by hand. Seeds with high market desirability may be grown in enclosed structures, either glass or plastic, where both growing conditions and pollination can be more adequately controlled.

Both flower and vegetable seeds are harvested from June to December. At the proper time, each seed-producing crop is cut and gathered into rows. For economical reasons, machinery is used for both operations whenever possible.

Each crop is then allowed a suitable length of drying time, one to three weeks, before it is threshed by machine. This is frequently done early in the morning before the dew evaporates from the crop in an effort to minimize seed loss, which can be substantial whenever the crop is extremely dry.

After being harvested, the seed is thoroughly cleaned and tested for germination by highly skilled registered seed technologists, who are regulated by germination standards set by the federal government for each species of seed.

Even though seeds may differ in their size, shape, color, and method of dispersal, they all have the same two basic components: (1) an embryo or infant plant and (2) a protective seed coat. They all are young plants in a state of suspended growth waiting for certain environmental conditions to be met before they can begin to germinate and grow.

The environmental conditions necessary for the germination of most seeds are:

SEED GERMINATION

1. An ample supply of water, which helps to soften hard seed coats, causes the embryo to swell and pierce its seed coat, and transports oxygen to the embryo.
2. A supply of oxygen, required of all life processes, especially the high rate of respiration associated with newly germinating seeds.
3. An optimum temperature range, which does vary with different seeds; i.e., early vegetable seeds are able to germinate in the temperature range of 35–45°F (2–7°C), whereas warm-weather crops require a temperature of 55°F (13°C) and above.
4. The absence of those materials or environmental conditions known to be detrimental to germination, such as
   a. Poisons
   b. High soil fertility levels
   c. Inhibitors.
5. An exposure to light, which is necessary for the germination of certain seeds. For others, light may not be required for germination, but can be the cause of seedling loss if the seed's planting depth is too great. For others, light actually prevents germination.

At maturity, the moisture content of most seeds is less than 10 percent, which causes them to remain dormant until sufficient moisture is absorbed. Because of the great differences between seeds, with respect to their individual storage and dormancy requirements, most gardeners depend upon the seed companies to treat the various seeds in the proper manner to ensure their germination when they are eventually sown.

Once the germination tests have been completed and the seed's viability has been confirmed, the seed must be subjected to its own unique storage and seed treatment demands to guarantee the germination rate anticipated by the home gardener.

To execute their function properly, the seed companies must know how to dry, store, treat, and properly package each seed type before making it available to the consumer. Because of the diversity of seeds, a variety of storage treatments are used. Most treatments require a minimum time span of a month and must be specific for the seed involved. The storage treatment employed for each seed is dependent upon a knowledge of why the seed remains dormant or fails to germinate.

Dormancy, or failure to germinate, can be temporary and can be caused by an unfavorable environment, or it can be more complex and involve one or more blocking mechanisms within the seed itself. These blocking mechanisms can be either physical, i.e., a hard seed coat, or chemical, involving the presence of inhibitory chemicals within or around the embryo. Tomato seeds, for example, will germinate only after being removed from the tomato and receiving a subsequent thorough washing.

The breaking of dormancy or afterripening is accomplished by a variety of treatments. The cause of the seed's dormant condition must be thoroughly understood and must be treated accordingly. Some treatments involve specific environmental conditions that enable the seed to fully mature and eventually germinate, while others are primarily concerned with attacking an impervious seed coat.

The changes that do occur within the seed during this period are believed to be the result of either the removal of growth inhibitors or the formation of growth-inducing substances. Experimental evidence has shown that the breaking of dormancy is associated with changes in pH, respiration rate, and enzyme activity.

To allow both water and oxygen to reach the embryo of a seed with a hard seed coat, the following treatments are employed:

1. Scratching the surface (This is the most common form of scarification; the next three are also variations of scarification.)
2. Use of boiling water.
3. Soaking overnight in water. (This also helps to remove certain germination inhibitors.)
4. Use of a dilute acid solution.
5. Alternate periods of freezing (32°F or 0°C) and thawing (60°F or 16°C).

Other seeds may require one of the following environmental treatments to break their dormancy:

1. Low temperature (40°F or 4°C), moisture and oxygen. Seeds are placed between layers of moist sand and/or peat moss, enclosed in plastic bags, and left in the refrigerator for one to six months (stratification).
2. Moist, low-temperature (33°F or 0.5°C) storage.
3. Dry, low-temperature storage.
4. Dry, high-temperature storage.
5. Alternating temperatures. Some seeds prefer temperatures between 40 and 80°F (4 and 27°C), while others require alternating temperatures of 60 and 105°F (16 and 40°C).
6. Soaking seeds in water and keeping them at temperatures just above freezing for several weeks (vernalization). This treatment has proved successful in allowing biennial plants to complete their growth in one year rather than the normal two years.

In addition to all the various treatments given to seeds to encourage them to germinate, most seeds should be treated with chemicals prior to planting to prevent seedling loss due to the presence of a damping-off fungus. There are several chemicals available that are designed to accomplish this feat. Many seed companies have performed this vital function for the gardener before their packaging is completed. Most seed packets, so treated, are labelled accordingly. For those packets not treated, one small pinch per seed packet is usually sufficient. The chemical used is added to the packet, shaken within the packet and the excess discarded.

Whenever a disease organism is assumed to be present within the seed itself, the seeds are placed in a cheesecloth bag and placed in a constant-temperature water bath of 120°F (49°C) for 30 minutes and immediately thereafter are sown.

After receiving its necessary period in storage and its specific pre-germination treatments, the seed is ready for packaging. A specified number of seeds are counted out, either by hand or by machine, into a dry, moisture-proof packet providing the eventual purchaser with sufficient seed to sow one row. Because of their delicate nature, great care must be exercised during the entire packaging procedure to prevent damage to the seeds.

Once the seeds have been packaged, they are kept cool and dry until they are distributed to the consumer. The seed companies may sell their seed either through their colorful catalogues, which they disseminate to interested gardeners, or on display racks found in both retail and wholesale outlets. Most seed packets have planting instructions and other important information clearly printed on the outside cover.

In addition to the production of flower and vegetable seeds for the home gardener, the other main divisions of seed production are:

1. Cereal grain and vegetable seed production for the large commercial farms primarily involved in wholesale food production.
2. Lawn grass seed primarily produced in the Pacific northwest, where high yields are obtained because of the climate and controlled growing conditions.

Chapter Nine

Planting

Prior to the planting of any plant, it is imperative that the grower be completely familiar with the correct planting technique involved to ensure its proper growth. Knowing when to plant is as important as knowing how to plant. Since planting is done both indoors as well as outside, and some of the techniques and considerations needed for success differ, each will be discussed separately.

INDOOR PLANTING

One definite advantage that plants grown indoors have over those grown outdoors is that their soil mixture can be prepared in advance to ideally match the plant with its soil preference. As stated earlier in the chapter on soils, the grower of outdoor plants usually has to accept the existing soil and must make every attempt to alter and condition it to provide the plants with a suitable growing environment. Drainage potential, pH, and organic content of the outdoor soil are three concerns which the grower must determine and correct, if necessary, before planting can occur.

Most indoor plants are planted in one of four basic soil mixtures. *For most plants and rooted cuttings*, a mixture of two parts loam, one part sand, and one part peat moss is ideal. *Acid-loving plants*, such as azaleas, gardenias, and camellias, prefer a mixture of two parts loam, two parts sand, two parts peat moss, and one part leaf mold or compost. Equal parts of loam, sand, and peat moss make the proper mixture for *ferns and begonias. Succulents and cacti* grow best in a mixture of equal amounts of loam and sand. Perlite or vermiculite can be used in place of sand.

To be successful in the preparation of any soil mixture, it is necessary to determine in advance the quantity needed, the best-suited mixture for the plants to be grown, and exactly how the mixing will be accomplished. Once the ingredients are mixed, the entire mixture must be thoroughly moistened before the all-important sterilization can be performed. The mixture is placed in a shallow metal pan and placed in the oven for 30 minutes at 200°F (94°C). If the odor resulting from the sterilization, actually pasteurization, is offensive, an enclosed charcoal grill can be used outdoors. It is advisable to wait one to two days after pasteurization before using the mixture. Any excess mixture left over after planting has been completed can be placed in any covered container and stored indefinitely.

For the germination of seeds and the propagation of stem and leaf cuttings, a sterile rooting medium is used. The media most preferred are sand, sphagnum moss, perlite, peat moss, and vermiculite. These may be used separately or as blends of two or more.

Once the soil mixture or rooting medium has been prepared, the next consideration is to select the proper container. Selection is usually based on the number of plants desired, the amount of space available, and the type of plant(s) and its method of growth.

The planting techniques used indoors are the following:

Potting. This technique is used to plant rooted cuttings, bulbs, tubers, corms, and rhizomes. With the exception of the rooted cuttings, the others are specialized underground stems, which when grown in pots must be positioned just below the surface of the soil mixture. Larger bulbs, like the narcissus, do not need to be totally buried beneath the soil, whereas the smaller bulbs must be. Tubers must be planted with their concave side facing upward, whereas corms must have their pointed side facing upward. Bulbs and rhizomes are much easier to plant because their root area is easily identified.

The pot selected for each plant type must be large enough to house the roots, yet small enough so that the plant never becomes dry in its container. Most rooted cuttings are initially planted in 2¼- to 4-in. (5.7–10-cm) pots, and are later transferred to larger pots as their roots grow and fill the pot.

The procedure used to pot a rooted cutting involves the following steps:

1. Place a small amount of the potting mixture in the bottom of the pot. *Note*: The positioning of crock material in the bottom of the pot is usually not necessary for small pots but should be used in containers 5 in. (12.5 cm) in diameter or larger. The exception to this rule: When potting rooted cactus cuttings, make certain that half the depth of the container is filled with crock material to allow for proper drainage.
2. Hold the cutting in one hand and place it in the center of the container, paying special attention to its height, so that at least the bottom inch (2.5 cm) of the stem will eventually be covered with the potting mixture.
3. While holding the cutting with one hand, use your other hand to carefully surround the roots of the cutting with the soil mixture until the pot is completely filled.
4. Use the fingers of both hands to press the soil mixture firmly around the base of the cutting. (Some horticulturalists prefer to use a potting

stick for this operation, but it is not necessary.) This procedure helps eliminate air pockets and also helps to keep the cutting in an upright position.

5. Make certain that the soil level is just below the rim of the pot so that the pot will be able to hold the water when it is applied.

6. Water the soil mixture thoroughly directly after potting occurs. This is done to prevent the presence of air pockets in the soil that would damage the roots. It is also done to prevent wilting of the newly potted cutting. Small quantities of water should be added periodically every few minutes until water flows freely from the drainage hole(s).

7. It is strongly suggested that each pot receive a second watering two to three hours after being potted, and be closely examined for the first week for signs of wilting.

Repotting. This is a practice done on a yearly basis during the spring months. The next larger container is selected if repotting is needed. To determine whether or not a plant requires repotting, it is first thoroughly watered, then inverted in your left hand with the main stem held by your fingers. The pot rim is tapped against a table, allowing the pot to be removed by your right hand. If roots are visible and thick outside the root ball, repotting is necessary. When repotting, place a small amount of soil in the bottom of the new container and then position the plant in the pot, making certain to achieve the proper depth and to keep the final soil level ½ in. (1.25 cm) below the pot's rim for subsequent waterings. The newly potted plant should receive liberal quantities of water directly after the repotting procedure has been completed. Water flowing through the drainage hole(s) gives evidence that the entire container has been well watered.

To help to ensure that the repotting procedure is successful, it is extremely important that the plant receive water every day for the first seven to ten days. It will take at least that long before the roots begin to venture into the new surrounding soil, and it is very possible for the small central earth ball to become dry while the new soil looks and feels wet.

Transplanting seedlings. Because of the small size of most seedlings, great care must be exercised when lifting them from their seed flat to avoid any damage to their small and fragile root systems. A dibble or pencil is used to make a deep hole for the roots of the seedling being transplanted. The seedling is held between the thumb and index finger or with tweezers and with great care the seedling's roots are placed in the hole. The thumb and index finger of the other hand are often used to pinch and depress the soil mixture immediately surrounding the seedling in an effort to minimize the danger of air reaching the roots and rapidly drying them out.

Once transplanted, the seedlings must be thoroughly sprinkled but not drenched. The first few days following transplanting are the most critical for the seedlings. They must never be allowed to dry out. If possible, they should be shaded from direct light for the first four to five days after being transplanted.

Planting bulbs for forcing (inducing them to flower before they would naturally outdoors). Smaller bulbs are planted below the soil surface, whereas the larger ones are potted so that the neck of the bulb protrudes slightly above the rim of the pot. The pots are placed in the garage or cold frame, where the temperature is about 40–45°F (4–7°C) for the prescribed number of weeks: tulips, 12; daffodils, 8–12; and hyacinths, 7. (See Chapter 31 for the complete details on the forcing of all bulbs.) The pots must be kept out of direct light and can not be forced into flower until both top and root growth are apparent. Tulips will be forced into flower in about five weeks, daffodils in about four weeks, and hyacinths in about two to three weeks.

OUTDOOR PLANTING

With the exception of permanent planters where the soil can be added, plants grown outside are dependent upon the existing soil. It is therefore important, before any planting occurs, that the grower fully understand the type of soil present, its texture, its pH, and then make the necessary adjustments and additions to the soil to insure optimum growth.

The two techniques employed in outdoor plantings are direct seeding and transplanting.

Direct seeding is used for lawns, vegetables, annuals, and perennials. It is important to prepare the soil before sowing any seeds. This involves turning over the top 6–8 in. (15–20 cm) of soil and thoroughly mixing into it the necessary soil additions. Once seeded, the seeded area must be well watered until the germinated seedlings are able to exist on their own. If the soil area is allowed to dry out, seedling growth may be either retarded or completely terminated.

The technique of *transplanting* is used to plant a variety of different types of plants. The ideal environmental conditions for transplanting are cool, cloudy days, with little or no wind and high humidity. If possible, every attempt should be made to take advantage of these conditions, which keep water loss to a minimum.

Figures 9-1 through 9-4 show, with explanations, how to transplant a tree or shrub properly.

1. Dig the planting hole 4–6 in. (10–15 cm) wider and deeper than the ball of earth.
2. Place soil and some form of decomposed organic matter in the bottom of the hole and mix thoroughly (Fig. 9-1).
3. Make certain that the top of the ball is level with or slightly below the existing ground level.
4. When the ball is in position, cut the twine surrounding the burlap and remove the burlap from the sides of the ball (Fig. 9-2).
5. Place topsoil around the ball and gently water the tree in place. Walking around the ball helps pack the soil and prevent air pockets.

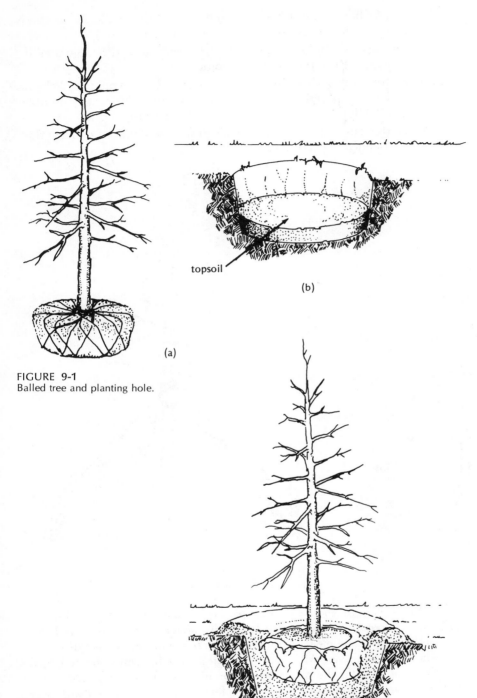

topsoil

(b)

(a)

FIGURE 9-1
Balled tree and planting hole.

FIGURE 9-2
Balled tree positioned in the hole.

6. Build an above-ground basin extending from the trunk to the outer limits of the ball to conserve and hold water (Fig. 9-3).

7. Prune some of the top branches to diminish the total foliage area in an effort to reduce the tree's evaporation rate.

8. Stake trees taller than 10 feet (3 meters). They should be held in place by guy wires attached to stakes for one growing season or longer. A minimum of three stakes is required (Fig. 9-4).

9. Wrap the trunk for the first year of growth to minimize damage due to environmental conditions.

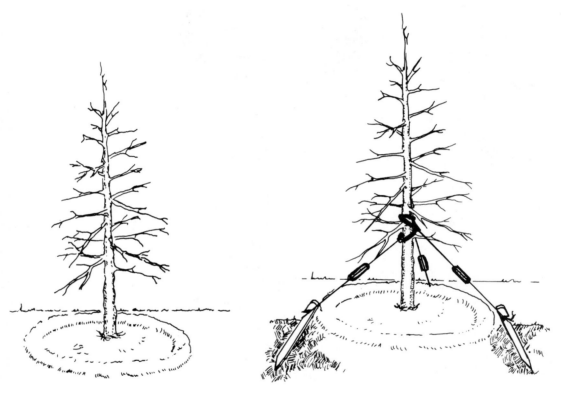

FIGURE 9-3
Properly planted tree.

FIGURE 9-4
Staking a newly planted tree.

When planting the foundation plantings from a landscape architect's plan, it is imperative to stake out the entire area and dig individual holes for each plant. A tape measure must be used to obtain the correct distances presented on the plan. All rocks and builder's debris must be removed from each hole. After the plants have been planted, the top 8 in. (20 cm) of the complete area should be worked up, leveled off, and covered with peat moss. The plants must receive ample supplies of water both at planting time and for the next few weeks until the plants become established.

Fruit trees and shrubs are transplanted in the same manner as trees are. Presented in alphabetical order are the other outdoor plant types and the method of transplanting them.

Annuals and perennials.   Before planting begins, you must first plan your garden, paying particular attention to the growth habits of each plant, such as its height, flowering time, and whether its growth habit is upright or spreading. New plants should be transplanted on cloudy days or late in the afternoon when the weather is clear.

Bare root plants.   The planting hole is prepared by forming a cone of soil in the center of the hole (Fig. 9-5). Spread the roots out over the cone's surface and cover them with loose soil, making certain that the roots do not all bunch together. Once positioned in the hole, the plant must be watered with a gentle stream from a hose. After the water settles, continue to fill the hole with soil and again gently water the plant.

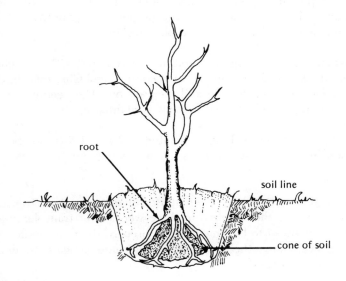

root

soil line

cone of soil

FIGURE 9-5
Cone of soil for bare root plants.

Berry fruits.   These should be planted as early as possible in the spring. To prevent them from drying out, they should be planted deeper than previously grown.

Bulbs.   For individual bulbs, a trowel or bulb planter can be used. For bed plantings, select the proper depth, and position the bulbs as desired. The depth for small bulbs is 3–4 in. (7.5–10 cm), whereas the larger ones are normally planted at a depth of 6–8 in. (15–20 cm).

Ground covers. These plants should be planted deeper than where previously grown to minimize water loss. Triangular spacing is employed as the most efficient method of distributing new plants (Fig. 9-6). Selection is based on height, light preference, and whether the foliage is evergreen or deciduous. The plants must be kept well watered for the first month after transplanting.

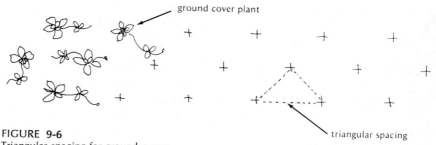

FIGURE 9-6
Triangular spacing for ground covers.

Hedges. Depending upon the species selected, it may be easier to dig a trench than to dig individual holes. The proper spacing between the individual plants must be known prior to planting time.

Reforestation. Small evergreen seedlings are planted with a specific purpose in mind, such as for timber, erosion control, and Christmas trees. The seedling is laid in a small hole and stepped on. The seedling is almost horizontal when planted, but will quickly straighten up.

Roses. The crown or swelling is planted 1–2 in. (2.5–5 cm) below the existing soil line. The distance between plants depends upon the variety selected.

Vegetables. Plants that have been started indoors must be transplanted with great care. For the first two to three weeks after transplanting, they should not be allowed to become dry. To prepare them for outdoor planting, some growers move their plants to an unheated garage to harden them off for four to seven days.

Chapter Ten

Garden Practices

There are a variety of garden practices used to maintain and beautify the home grounds. It is equally important to know why they are done as well as to know how to perform each one. The incorporation of the following practices into the overall gardening scheme should help to guarantee a successful and picturesque landscape.

Edging is done to separate one growing area from another, e.g., the lawn from the garden. It is frequently used to contain or restrict the growth of a specific plant whose presence in another area of the garden would be undesirable. When done properly, edging gives the added dimension of a neat, well-kept landscape. The materials used in edging are varied and include: *EDGING*

1. *Edging iron.* It is half moon in shape at the end of a long handle and is used solely to restrict the growth of grass.
2. *Edging spade.* It is a rectangular-shaped shovel on a short handle. It serves the same function as the iron, but may also be used for some planting.
3. *Bricks, stones, or concrete.* These construction materials are formed into low walls which define one growing area from another.
4. *Metal strips.* Usually made of aluminum, these come in a variety of widths. They may be left visible above the ground or buried completely beneath the soil to restrict the growth of a plant that sends its rhizomes out below ground as its means of expansion.
5. *Living plants.* Certain low-growing plants are used to separate one growing area from another. Boxwood and dwarf heath are two that may be used.

It is a well-known fact that all green plants require fertilization for optimum growth. The quantity of fertilizer dispensed and the methods of application are two concerns that may not be completely understood. The amount of fertilizer added at any time is based on the following:

1. Knowing the growth habit of each plant and its fertilizer requirements.
2. Realizing that less is required during periods of slow growth, i.e., winter.
3. Following the manufacturer's recommendations on the fertilizer package.

Fertilizer is added to plants either in granular form or in liquid form. In granular form, it can be applied either by hand or by using a spreader.

When one is applying fertilizer by hand, the best method of application is called the figure eight method, where the fertilizer is dispensed through the fingers while the hand is making a figure eight pattern. This method is fine for a limited area, but on a larger scale, the chance of both burning plant foliage and missing areas is great.

For large areas of lawn and garden, a spreader is more practical. The application is usually uniform, unless the drive wheel on the spreader slips, thus causing an area to be missed. It is advisable to turn off the spreader at the end of each run, since there is a danger of excess at the turns.

In liquid form, fertilizer can be applied by using a sprinkling can when the area is small, and a hose applicator if the area is large. The hose applicator works on a siphon principle, and there is no immediate guarantee that an application has been uniform. This method can also be used to apply a foliar spray directly to the leaves of the plant.

LAWN MOWING
Most individuals take this procedure for granted. When the lawn appears less than ideal, many different reasons are usually given for the problem with little or no thought to how and when the lawn was cut. The tips for success for both new and established lawns with respect to proper mowing procedures follow:

New Lawns

1. Set the mower height at 1½–2 in. (3.75–5 cm) for Kentucky bluegrass lawns. *Note:* The height will vary for other turf grass types.
2. Cut the new seedlings when they reach a height of 2 in. (5 cm).
3. Mow when the growing surface is dry.
4. Use a grass catcher or lawn sweeper to collect the grass clippings.
5. Do not allow fallen leaves to collect on the new grass seedlings, since they will smother the grass.
6. Make certain that the blade of the mower is sharp.

1. Do not cut Kentucky bluegrass too short or too soon in the early spring, since the root system may be damaged.

2. Mow at regular intervals of time, such as once a week.

3. Cut the grass when its height is approximately ½ in. (1.25 cm) greater than the mower's cutting height.

4. Whenever the grass gets too tall, reduce its height a small amount at a time with consecutive mowings.

5. Decrease the problems related to both disease and weed infestation by collecting grass clippings during the mowing procedure.

Established Lawns

The technique used for this practice is to pinch off the terminal growing tip, once the plant has reached a certain size. Some plants are pinched when they become 3–5 in. (7.5–12.5 cm) tall, whereas others are pinched when they produce three to four sets of true leaves. Pinching is done for the following reasons:

PINCHING BACK

1. To force the plant to develop branches.

2. To increase the number of buds.

3. To obtain a compact, bushy plant rather than a tall, leggy one.

4. To encourage both flower and fruit development in certain plants, while retarding the same development in other plants.

Annuals that benefit from pinching are ageratum, browallia, calendula, chrysanthemum, petunia, phlox, pinks, snapdragon, verbena, and zinnia. Those annuals that are harmed by pinching are cockscomb, impatiens, poppy, and stock.

Raking is done to accomplish the following functions in the garden:

RAKING

1. To establish the proper grade necessary for drainage off any growing surface.

2. To prepare the soil for planting by leveling off the surface.

3. To introduce grass seeds to the newly prepared seed bed just below the ground surface.

4. To keep the landscape clean and neat in appearance. Raking is used to clear virgin ground, and to remove leaves, thatch, and grass clippings from the lawn and garden.

Root pruning is a technique that must be familiar to any gardener who plans to move any existing plant from one location to another one. The same technique is employed, no matter whether the plant is already on the property or is growing wild in the woods.

ROOT PRUNING

This technique usually extends over two growing seasons. In the fall of the year, a circular trench, 18 in. (46 cm) deep, is dug around the plant, at a distance of 18–24 in. (46–61 cm) from the trunk for most plant material easily handled by most gardeners. The trench can range in width from three to six inches (7.5–15 cm), and should be cleared of rocks and existing roots.

During the winter months, the plant will develop new roots within the remaining soil ball, thus replacing those lost during the fall root pruning.

In the spring, the plant can be easily dug up and transplanted to its new location. The digging must be done before the buds enlarge and burst open. It is advisable to prune some top growth when the plant is dug, to help balance top growth with root growth.

SEEDING

Since some seeding is done indoors in preparation for the transplanting of seedlings later outdoors, the important considerations mentioned here will include both indoor as well as outdoor seeding:

1. The proper time to sow the seed for each plant, and which plants do better when seeded indoors must be known.
2. The properties of your soil must be understood. If it is predominantly a heavy clay that bakes as it dries out, both sand and compost must be added.
3. The soil surface for any seedling should never be allowed to dry out.
4. For proper germination of certain seeds, shading may be required.
5. Indoors, most seeds germinate faster when placed in a warm, dark space. Covering the entire container with moist newspaper helps to keep the area dark and to keep water loss to a minimum. Once germination occurs, the newspaper must be removed.
6. In the seeding of new lawns, measurements play an important part. For purely economical reasons, it is imperative to follow the directions on the seed packet.
7. Seeds with a hard outer coating may require a thorough soaking prior to sowing to enable the seed to penetrate through this coating.
8. Once germination has occurred and the new seedlings have formed two to three sets of true leaves, fertilizer in liquid form must be added to ensure the continuous growth of the seedlings.

STAKING

Too often, this important procedure is either completely ignored or considered unimportant. The reasons that staking should be done are the following:

1. To provide support for weak plants unable to stand on their own.
2. To insure protection against the wind and severe storms that may flatten certain plants.

3. To encourage upright growth in certain plants, e.g., tomatoes.
4. To prevent newly transplanted trees and shrubs from being up-rooted, because of the freezing-thawing conditions that exist during the winter months.

A wide range of different types of structures can be employed to provide plants with the proper support. They are:

1. walls
2. trellises
3. fences
4. arches
5. stakes placed vertically in the ground
6. pergola—a trellised arbor
7. metal hoops on a stake
8. wire stretched between posts
9. espalier
10. teepee, using three poles
11. wall nails, cloth or leather straps—building attachment
12. metal frame lattices
13. stakes, wire, and pieces of rubber hose to prevent larger trees from being uplifted by the wind—a minimum of three stakes, equally spaced, is required.
14. circular baskets made of reinforcing steel for tomatoes—minimum distance should be 18 inches

A severe buildup of grass clippings from successive mowings is known as *thatch*. It must be removed periodically from the lawn, or the grass will suffer. Thatch is removed from the lawn surface by using a mechanical thatcher, rented from a power tool supplier, or by using a metal leaf rake held in an upright position at right angles to the lawn and by giving the lawn surface a stiff raking. It is advisable to remove thatch only during periods of active growth so that the lawn grasses have sufficient time to rebound before certain environmental conditions suppress the growth. If done in the spring, it must be done prior to the application of any preemergence crabgrass killer.

THATCH REMOVAL

Chapter Eleven
Pruning

Pruning is a specialized garden practice often ignored or relished by the ignorant. In either case, the plant's pruning requirements suffer from neglect or overindulgence.

Pruning is defined as the systematic removal of plant parts with a definite purpose in mind. The reasons for pruning are:

1. To restore proper balance between root and top growth at the time of transplanting.
2. To remove injured or dead parts.
3. To improve production; i.e., flowers, fruit, or timber.
4. To control size and attain the desired appearance.
5. To improve or decrease shade, whichever is desired.
6. To allow certain plants more growing space by thinning out other plants; e.g., timber production.
7. To rejuvenate older plants.

To accomplish the varied techniques of pruning every gardener should own the following tools:

1. A hand pruner for small branches with a maximum size about that of a pencil [Fig. 11-1(a)].
2. Lopping shears to remove branches from pencil size to 1¼ in. (3.2 cm) [Fig. 11-1(b)].
3. A pruning saw for branches that exceed 1½ in. (3.8 cm) in diameter [Fig. 11-1(c)].
4. Pruning or hedge shears for trimming hedges and certain evergreens [Fig. 11-1(d)].

Additional equipment for the gardener would be a pole pruner to remove branches up to 20 feet (6 meters) in height and tree wound paint to be used on cuts larger than ½ in. (1.25 cm) in diameter.

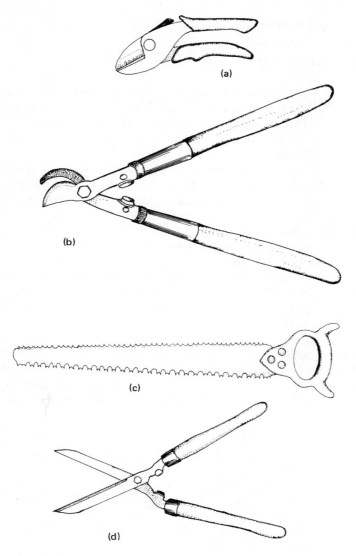

FIGURE 11-1
(a) Hand pruner. (b) Lopping shears. (c) Pruning saw. (d) Hedge shears.

For the gardener to be successful, it is imperative that all pruning tools should be

1. Of good quality.
2. Sharp and clean.
3. Handled with care and respect.
4. Properly tightened.
5. Never forced beyond their capabilities.

6. Selected to fit the job at hand.

7. Never twisted while being used.

8. Oiled at frequent intervals.

Before you start to prune any tree, shrub, or hedge, you must realize that pruning requires both finesse and thought rather than simply ambition. You must be thoroughly familiar with the growth habit of the plant you intend to prune. Too often, the natural shape of the plant may be destroyed and lost forever.

Every gardener must learn the difference between a vegetative bud, which is flat and pointed, and a flower bud, which is round and fat. Removal of a large number of flower buds will result in a decrease in flower or fruit production. It must also be understood that both types of buds are formed during the summer months.

Plant stems should be pruned just above a visible bud that points away from the center of the plant. Each cut should be made on a 45-degree angle (Fig. 11-2). The angle helps protect the cut stem from damage caused by water and pest invasion. All branches that crisscross within the plant should be removed. These spoil the plant's appearance and limit the desired growth.

Thinning out is designed to open up the plant so that a more natural look can be maintained. It is accomplished by the removal of small branches back to either a side branch or the main trunk. This procedure requires few cuts and thus results in fewer apparent pruning stubs.

Heading back, designed to increase the bushiness of the shrub

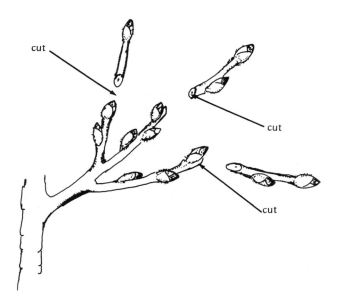

FIGURE 11-2
Stem pruning.

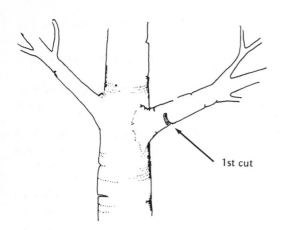

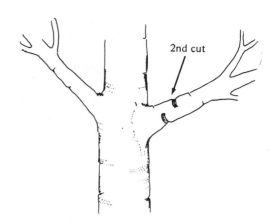

FIGURE 11-3
Limb removal—first cut.

FIGURE 11-4
Limb removal—second cut.

while still maintaining the natural shape of the plant, is performed by cutting a branch back to a bud or a side branch.

The removal of large tree branches must be accomplished in three steps. The first cut is an undercut 8–10 in. (20–25 cm) out from the final cut. This undercut should go one-third of the way through the stem (Fig. 11-3). This first cut is done to remove some of the weight of the branch being cut and also to prevent removal of bark beyond the final cut.

Make the second cut down through the stem to meet the first cut. The second cut should be slightly farther out on the branch than the first cut (Fig. 11-4). This procedure reduces the danger of injury to the tree and to yourself. Make certain that you are positioned properly when making the second cut so that the limb does not strike you or knock you off a ladder.

The final cut is made as close to the trunk as possible to both speed

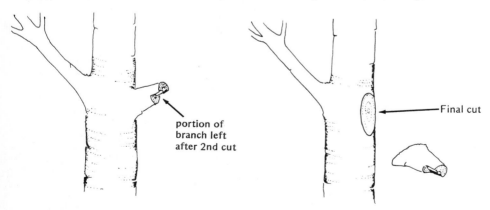

FIGURE 11-5
Limb removal—final cut.

79

the recovery period and reduce the possibility of attack by either insects or disease organisms.

The balance of this chapter explains the proper time and procedures used to prune a variety of plant material.

Hedges are trimmed to obtain a desired height and shape. They may require more than one pruning during the year. Only the new soft growth should be clipped.

Hybrid tea *roses* must be pruned each spring, cutting each cane back to within 6 in. (15 cm) of the ground and leaving three to four buds on each cane. Climbing roses and the hedge type are pruned by simply removing only the dead wood.

Spreading *evergreens* are pruned when these plants are producing soft new growth. Some gardeners leave the branches longer than they should so that they have a source of greens at Christmas time.

Nonflowering shrubs and trees are pruned to remove the dead wood and to maintain the proper shape. Pruning deciduous plants during the dormant season helps the gardener determine more easily which stems should be removed.

Shrubs and trees that flower before July 1 should be pruned directly after flowering; those that flower after this time should be pruned in the early spring.

Vines should be pruned in the same manner as described above for flowering shrubs.

The pruning of fruit trees is more involved and requires specific procedures during the life of the tree. At planting time, the main trunk is already well defined, being 20–24 in. (50–60 cm) high. The main leader or central branch must be removed. This procedure may be already accomplished by the nursery. The remaining branches should be reduced to no more than five and these will become the main branches of the tree, referred to as *scaffolds* (Fig. 11-6). It is important to make sure that these five branches are healthy and form an angle of 30 to 45 degrees with the trunk.

FIGURE 11-6
Pruning a fruit tree at planting.

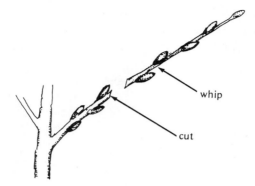

whip

cut

FIGURE **11**-7
Pruning a whip.

As growth progresses, the main branches should be allowed to extend upwards and outwards for about 6 ft (1.84 m), giving rise to the framework branches. These framework branches should extend upward another 4 ft (1.22 m), and be allowed to branch into a profusion of smaller branches which will carry the fruit.

At fruit-bearing age, apples, cherries, pears, and plums are pruned in the following manner. Last year's new growth, called *whips*, is pruned to one-third of its total length; each should contain four to five buds (Fig. 11-7). On the remaining portion, the lower buds will develop into new growth, called *spurs,* which will carry the fruit, while the top bud forms a new whip.

When peach trees reach fruit-bearing age, half of the length of each branch must be removed to allow for proper development of the fruit, since the fruit develops on last year's growth. When growth is excessive, complete removal of some branches may be necessary.

Espalier and topiary are two forms of pruning which allow the gardener to be creative as well as artistic in his pruning efforts.

Espalier pruning is accomplished by growing the desired plant flat against a suitable support structure, such as a fence, building, wall, or trellis. Plant material must be selected that will be conducive to this form of pruning, and it must be realized that the desired final effect may not be attained for several years because of the constant work involved, i.e., the training and tying of the individual branches in their proper location, coupled with the pruning.

Topiary is the art of trimming specific trees and shrubs into different geometric shapes. Its artistic possibilities are endless, and it requires the same type of diligence, patience, and pruning skills associated with espalier pruning.

SPECIALIZED TYPES OF PRUNING

Chapter Twelve

Garden Measurements

The importance of garden measurements must never be taken lightly. Accurate measurements are used to avoid any awkward appearance in the landscape caused by plantings that are either too close or too sparse.

Since successful gardening depends upon the administration of a variety of chemicals in the correct concentration, improper measurements could result in the destruction of plant material because of the toxic quantities added. Proper measurements will prevent the addition of excessive amounts of fertilizer, lime, and pesticides, which is undesirable for purely economical reasons.

The measurements made in the total landscape are either *linear*, dealing with straight line determinations, or *area*, involving a two-dimensional plot.

LINEAR MEASUREMENTS

Linear measurements are made either in inches (centimeters) or in feet (meters). A tape measure is most commonly used for all landscape plantings, such as the planting depth, the distance between plants, the distance from the house, and the distance between fence poles.

In commercial operations, correct spacing between plants is imperative to ensure maximum production. For most plants, this has been experimentally tested and proven.

When one is planting certain vegetables in rows, a marked board, measured in inches (centimeters), that can be placed on the ground near the row is commonly employed. Most gardeners use all four edges of the board and mark each edge with a different spacing. The four spacings most commonly used, one per edge, are 3, 4, 5, and 6 in. (7.5, 10, 12.5, and 15 cm). The normal spacing for peas planted in rows is 3 in. (7.5 cm) with a distance of 18 in. (45 cm) between rows. The board, with markings at 3-in. (7.5-cm) intervals, is used to ensure even distribution of the seeds. Three units of the edge with 6-in. (15-cm) markings are used to obtain the correct space between the rows.

To obtain a rough approximation of a long linear distance, a walking pace can be used. The distance travelled in two walking steps is approximately the same as a person's height.

Area measurements are determined by multiplying the width of a plot by its length and are expressed in square feet (square meters). The addition of the proper amount of all soil chemicals, grass seed, and mulches is dependent upon knowing the exact area to be treated.

It is recommended that homeowners know the exact measurements of all their lawn and garden areas. Since these dimensions seldom change throughout the years, an initial measurement of these areas should be made and recorded. Since many of the manufacturer's labels give their directions based on the amount applied to an acre, the chart in Fig. 12-1 can be used to compute any surface area, be it large or small, in acres.

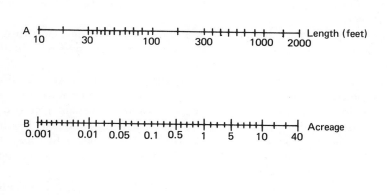

FIGURE 12-1
Chart to compute surface area.

First, measure the length and width of the area in question. Column A is used for length and column C for the width of the area. To determine the actual acreage of the measured plot, draw a straight line from the dimension in column A to that in column C. Where the line crosses the center column (B) is an approximation of the acreage of the land being measured. For a plot 200 ft (61 m) by 40 ft (12.2 m), the acreage would be about 0.2 (two-tenths) acre. If the recommendation is to apply fertilizer at the rate of 500 lb (227 kg) per acre (0.4 hectare), this plot should receive 100 lb (45 kg).

To sow 1000 ft² (92 m²) of soil area into lawn, the amount of grass seed required depends upon the variety of lawn desired. To sow this area with Merion bluegrass, only 2 lb (0.91 kg) of seed would be required, as compared to 4 lb (1.82 kg) of Kentucky bluegrass and most popular mixtures containing the bluegrass. To cover the same area with either perennial ryegrass or the fescues would require 3–6 lb (1.35–2.7 kg) of seed.

Five pounds (2.27 kg) of ground limestone spread evenly over 100 ft² (9.2 m²) of ground surface will raise the soil pH by three-fourths of a unit. This is equivalent to one ton per acre (900 kg per 0.4 hectare). To treat the same 40-by-200-foot area previously mentioned would necessitate the addition of 400 lb (180 kg) of lime. Ground limestone is sold to gardeners in 50-lb (22.5-kg) bags, and therefore eight bags would be needed to treat the area.

For most soil surfaces where grass is grown, the usual rate of fertilizer to be applied is one pound (0.45 kg) of nitrogen per 1000 ft² (92 m²). To determine this amount, divide 100 by the percentage of nitrogen in the fertilizer. For example, when a 10-6-4 fertilizer is used, 10 lb (4.5 kg) of the fertilizer is needed.

When the decision has been made to use sod to cover a hard-to-seed area, it is important to realize that the sod is generally sold in rolls that measure 18 in. (45 cm) by 6 ft (1.84 m).

For most mulches and granular herbicides (weedkillers), the directions on the package must be followed. Spreading the material too thin will not satisfy the manufacturer's recommendation and will usually give results that are less than ideal. Also, using the old mistaken belief that doubling the recommended amount will be twice as effective is poor policy. In some cases, this is simply poor economics, while in other situations actual harm can be done to plant material.

Most pesticides, i.e., insecticides, fungicides, miticides, herbicides, and rodenticides, and some fertilizers are sold in concentrated form with directions concerning the amount needed for proper control. Many times, the directions are stated per acre or per 100 gallons of water, leaving the person with a smaller area totally bewildered. Table 12-1 should prove helpful to any gardener.

Table 12-1
CONVERSION TABLE

Concentrated form	Label statement	Use
Solid	1 lb/100 gal water	1 tbsp/gal water/(3.75 liters)
	2 lb/acre	2 tbsp/gal water
Liquid	1 pt/100 gal water (0.56 l/375l water)	1 tsp/gal water (1.3 ml/liter)
	1 qt/acre (1.12 l/0.4 hectare)	2 tsp/gal water (2.6 ml/liter)

All gardeners should be acquainted with the following three equalities associated with applications of chemicals to specific soil areas:

1. 1 oz (28 g)/100 ft^2 (9.2 m^2) = 27.2 lb/acre (12.3 kg/0.4 hectare)
2. 1 pt (0.56 l)/1000 ft^2 (92 m^2) = 5 gal/acre (18.75 l/0.4 hectare)
3. 2.5 lb (1.14 kg)/1000 ft^2 (92 m^2) = 100 gal (375 l)/acre (0.4 hectare)

Most of the pesticides sold commercially are viable for years when stored in a dry location. Solutions can be made from these pesticides in gallon containers and the excess can be stored for future use, but most gardeners prefer to use fresh material each time they spray even though the number of plants to be treated is small. When the quantity of solution is less than a gallon or if the manufacturer's direction are given in either tablespoons or fluid ounces, Table 12-2 should enable the gardener to make any solution that is required.

Table 12-2
IMPORTANT CONVERSIONS FOR THE GARDENER

1 teaspoon	= $^1/_3$ tablespoon	= $^1/_6$ fluid ounce	= 5 milliliters
3 tsp	= 1 tbsp	= $^1/_2$ fluid oz	= 15 ml
16 tbsp	= 1 cup = 1 3-in. clay pot	= 8 fluid oz	= 240 ml
2 cups	= 1 pt	= 16 fluid oz	= 480 ml
2 pt	= 1 qt	= 32 fluid oz	= 0.96 l
8 cups	= 2 qt	= 1 6-in. clay pot	= 1.9 l
4 qt	= 1 gal	= 3.8 l	

The directions printed on certain pesticides suggest a specific dilution rate from the concentrated form. In most instances, Tables 12-1 and 12-2 should be sufficient to solve most of the dilutions that the gardener must face. However, the following dilutions, based on the number of tablespoons per gallon (3.8 l) of water may be helpful.

¼ tbsp in 1 gal makes a 1:1000 dilution.
1 tbsp in 1 gal makes a 1:250 dilution.
2½ tbsp in 1 gal makes a 1:100 dilution.

Chapter Thirteen

Insect and Disease Control

All growers of plant material should maintain a definite program of insect and disease control which encompasses the following methods:

1. Growing plant varieties and hybrids that are resistant to disease.
2. Purchasing and growing only healthy plants.
3. Keeping all growing areas clean and sanitary.
4. Spraying and dusting with pesticides on a specific schedule.

In addition to these methods of control, most growers should be able to diagnose the cause for any substandard growth or injury to the plant. It is important to know the specific cultural requirements for each plant grown and what to expect of the plant at different times during the year. One plant may flower only during the winter months and go dormant directly after flowering, while another plant may flower continuously except during periods of low light intensity.

LEAF DIAGNOSIS Careful examination of the plant leaves may indicate to the grower what type of problem is present. For example:

1. If a portion of the leaf area is missing, *chewing insects* are suspected.
2. If the leaves appear yellow and deformed, *sucking insects* are probably present.
3. When the leaves appear discolored, *nutrition* or improper fertilization, not pests, is indicated.
4. The presence of spots on the leaves usually suggests the presence of *disease organisms*.
5. If the plant shows partial collapse, some *wilt organism* may be present.

To keep a large number of house plants free from injury, whenever a specific insect or disease organism has been identified, an application of a general-purpose spray containing the following ingredients should be applied to the plants in any well-ventilated area outside the house:

Measure	Ingredient	Employed for
1½ tsp	50% malathion (emulsifiable concentrate)	insect control
1 tsp	Kelthane	spider mite control
½ tsp	household detergent	proper blending of the other ingredients and adequate leaf attachment
2 tbsp 1 gal water (3.77 liters)	50% Captan	disease control

The three chemicals listed above can be purchased individually in concentrated form and mixed whenever needed. In the long run, the expense involved in the formation of the spray will be far less than the purchase of prepared solutions. These same ingredients, in larger amounts, can also be used to treat outdoor plants. A Windex spray bottle can be used to spray indoor plants.

The diseases attacking plants are divided into two general categories: *DISEASES*

1. Those caused by bacteria and fungi.
2. Those caused by a virus.

To minimize the amount of disease control, the grower must insure healthy plant growth by providing the proper environmental conditions. Because disease organisms are microscopic in size, the grower must look for telltale symptoms on the plant. Once a disease organism is discovered and identified, a suitable fungicide is applied to the plant to control it. Table 13-1 is a generalized listing of plant diseases, giving the description and control of each disease.

Table 13-1
PLANT DISEASES

General name	Description	Control
Canker	A localized dead area (found primarily on trees)	Small ones are removed by cutting out the infected tissue. Keep plants healthy and well fertilized.
Gall	An abnormal growth or swelling	Many galls are removed by pruning out infected tissue; others are controlled by a fungicide.

Table 13-1 (continued)
PLANT DISEASES

General name	Description	Control
Leaf blight	Discoloration or destruction of leaf tissue	Destroy all available leaves each fall; spray plants every two weeks with a fungicide.
Mildew	White areas on the leaf; *powdery* (on top) *downy* (on the bottom)	Destroy all infected plant parts; spray with sulfur fungicides (may harm vine-grown vegetables).
Rot	A disintegration of any plant tissue	Control root rot by crop rotation. Stem and fruit rot are usually kept in check by sanitation and spraying with a fungicide every two weeks.
Rust	Appearance of bright orange spores on both leaves and fruits	Control may involve removal of an intermediate host. Select rust resistant varieties and spray with a fungicide.
Wilt	Wilting due to the presence of a disease and *not* to a lack of water	Prevention involves dusting the soil, spraying with a fungicide, planting resistant varieties, and crop rotation.
Virus diseases	*The plant leaves may:* 1. Become distorted and malformed. 2. Become pale in color. 3. Become mottled, displaying a mosaic effect. 4. Curl or roll up. 5. Appear stunted in size.	Prevention involves a rigorous sanitation program requiring that all infected plants be completely destroyed.

Treatment of plant diseases involves the spraying or dusting of infected plants with a suitable fungicide. To minimize the danger of a possible infestation, growers of plant material frequently examine their plants for symptoms and follow a routine spray schedule, usually once a week. Table 13-2 is an alphabetical listing of available fungicides and related comments about each one.

Table 13-2
COMMON FUNGICIDES

Name	Comments
Bordeaux mixture	A copper sulfate and lime mixture used for blights.
Cadmium chloride	A 10% solution is used as a lawn fungicide.

Table 13-2 (continued)
COMMON FUNGICIDES

89

Insect and
Disease
Control

Name	Comments
Captan	A general multipurpose fungicide.
Dinocap (Mildex, Karathane)	Used to control powdery mildew—an additional benefit is suppression of the mite population.
Dyrene	Used to control most lawn diseases.
Ferbam	A general protectant fungicide; may be undesirable because of its black residue.
Folpet	Used for leaf spots, blights, and powdery mildew.
Maneb	A multipurpose fungicide.
Sulfur	Used alone for powdery mildew; when mixed with lime, it becomes a suitable fungicide and insecticide.
Zineb	A general protectant fungicide.

Note: The directions on each container must be carefully followed.

INSECTS

The insects causing damage to plants are usually classified according to their method of injury and are placed in one of three groups:

1. The chewing ones.
2. The sucking ones.
3. Those insects which transmit plant diseases on their bodies.

Knowledge about the stages in the life cycle of a specific insect, i.e., from egg to adult, and its method of injury are essential before the grower can effectively control it. The life cycle of many insects involves four stages: egg, larva, pupa, and adult. These insects go through a complete life cycle, while others have an incomplete life cycle, in which they go through only three stages: egg, nymph, and adult. The nymph replaces both the larva and pupa stages.

Until recently, insect control was accomplished in the following three ways:

1. *By cultural methods.* Using cleanliness, crop rotation, and pruning, and by planting resistant varieties.
2. *By mechanical methods.* Using traps, paper collars, and tar paper discs to catch or deter pests. Hand picking is also used.
3. *By chemical methods.* Spraying or dusting the plants with insecticides.

The chewing or biting insects are controlled by applying a *stomach poison* directly to the plant. The sucking insects are controlled by spraying them directly with a *contact poison* or by adding to the soil a *systemic insecticide* which is absorbed by the plant roots and becomes part of the plant's sap. The third group is held in check by spraying with either contact or stomach insecticides as an insurance against possible disease infestation.

Recent attempts at achieving insect control have pursued the following avenues of approach:

1. *The radiation of adult males.* This procedure is done to male insects in captivity and renders them sterile. They are then released at the proper time to mate with their female counterparts with the net result that no fertile eggs are produced.

2. *Interruption of the life cycle.* If this can be done by chemically treating the insects while they are in one of their immature stages, the total insect population will be decreased and fewer adult insects will be found. Since many insects overwinter in the soil, the application of an insecticide to the soil when the insect is in the larva or pupa stage should be very effective at decreasing the total insect population.

3. *The use of beneficial insects.* Lady bug beetles and praying mantis are two insects that consume harmful insects. Certain seed catalogues now supply their clients with either the actual insects or their eggs as a means of insect control.

4. *Using beneficial plants.* Certain plants are grown in close proximity to more desirable plants in an attempt either to lure attacking insects away from the valuable plants or to create an environment undesirable to the various invading pests.

Table 13-3 lists the common insects found on plant material, and related comments about their habits and their control.

Table 13-3
COMMON INSECTS

Type Name	Comments and Control
Chewing Bagworms	Overwinter in bags primarily on evergreen and deciduous trees. Control by picking off the bags in the winter or spraying with malathion or Carbaryl.
Beetles	Harmful to plants in their grub, larval, and adult stages. Controlled by applying a stomach poison to the plant or a soil treatment with diazinon.
Borers	Attack woody plant stems in the form of grubs, caterpillars, or larvae; may overwinter in infected twigs and old vegetable garden remains. Control is sanitation and spraying plants during active growth with malathion.

Table 13-3 (continued)
COMMON INSECTS

91

Insect and
Disease
Control

Type Name	*Comments and Control*
Cutworms	Overwinter in the soil or in garden debris; in the spring, the larvae cut off seedlings at ground level (cabbage, tomato, and flower seedlings). Use any suitable stomach poison, applying it both to the seedlings and the surrounding soil.
Slugs	Chewing insects that attack almost all plant material. Control by using Metaldehyde, lime, saucers of beer, and wooden planks that will attract them during the day so that they can be easily found and destroyed.
Sucking Aphids	Light green insects found on flower buds or young terminal shoots. Spray with malathion.
Cyclamen mites	Invisible insects causing twisted stems, stunted plants, misshapen foliage, and few and imperfect flowers; able to crawl from plant to plant only if the leaves touch. Spray with Kelthane.
Lace bugs	Have lacelike wings and are found on the underside of the leaf, where they cause a discoloration. Spray with malathion.
Leaf nematodes	Live in the soil and are microscopic; cause a yellow-brown, pie-shaped discoloration, which may cause the entire leaf to become brittle and drop off. Spray with a nematocide.
Mealybugs	Cottony white masses found in the leaf axils; secrete a sticky substance on the leaves. Spray with malathion.
Red spider mites	May be red, green, or black in color; cause leaves to turn yellow and appear cobwebby; leaves may also become curled and twisted. Control is by syringing with water and spraying with Kelthane.
Scale	Appear as hard discs on both stems and leaves. Spray with malathion.
Thrips	Leaves show small brown spots underneath, and may curl and display papery scars; flower buds drop, while flowers become deformed and discolored. Spray with malathion.

Table 13-3 (continued)
COMMON INSECTS

Type Name	Comments and Control
White flies	Insects that suck the underside of the leaf; fly when the plant is shaken. Spray with malathion.

For either an insecticide or fungicide to be effective and sold to the public, it must do the following:

1. Give quick and effective results.
2. Not harm plant foliage.
3. Adhere satisfactorily to plant parts.
4. Not deteriorate when stored.
5. Not be toxic to humans or other animals.
6. Be inexpensive.
7. Not harm the environment.

Table 13-4 gives an alphabetical listing of the currently available insecticides, what insects each controls, and pertinent comments about each.

Table 13-4
AVAILABLE INSECTICIDES

Name	Controls	Comments
Bacillus thuringiensis	caterpillars	A safe *microbial* insecticide.
Carbaryl (Sevin)	most insects	Has replaced many of the more toxic insecticides.
Cryolite	chewing insects	Used on fruit and shade trees.
Diazinon	most insects	Suitable replacement for chlordane.
Dimethoate	most insects	Least toxic systemic insecticide; may damage certain flowering plants.
Dormant oil	scale, aphids	Used on fruit trees.
Kelthane	spider mites	Should not be mixed with sulfur or lime.

Table 13-4 (continued)
AVAILABLE INSECTICIDES

Name	Controls	Comments
Malathion	most insects	Mixes well with other chemicals; controls both aphids and mites.
Metaldehyde	slugs, snails	Compound placed directly on soil.
Methoxychlor	most leaf feeders	Has a low toxicity to humans.
Pentac	spider mites	May cause foliage damage.
Pyrethrum	most sucking insects	Made from flower heads of the composite family; found in many all-purpose insecticides.
Rotenone	most insects	Nonpoisonous to humans; used by vegetable growers. Loses its effectiveness rapidly.
Sulfur	mites	A wettable powder type is used.
Tedion	spider mites	A slow-acting, but long-lasting insecticide.

Chapter Fourteen

Landscape Planning

Landscape planning involves a wide range and degree of activity that may involve a simple planting of trees or shrubs performed by a gardener or maintenance contractor or a complete landscape project requiring a designer, detailed plans, carpenters, masons, and nurserymen to accomplish the desired objective.

Regardless of the simplicity or complexity of the project, an individual with design ability and practical knowledge is required to plan the project to insure the desired results. Some homeowners are avid gardeners with a wide knowledge of plant material and may know what plants they want incorporated into a particular area.

Others may have good design concepts, but may possess little or no knowledge of plant or constructional materials and may require the necessary practical knowledge associated with these materials. The majority of clients, however, with whom the landscape designer usually works do not possess the necessary knowledge of design or materials and are, therefore, dependent upon his services.

The extent to which the designer will be able to assist the client will depend upon the designer's natural ability, his schooling in landscape design, and his actual experience.

EDUCATIONAL
CONSIDERATIONS

If an individual is primarily interested in working with plant material, the horticultural and design courses available at most state colleges provide the necessary information. Many of these courses provide detailed information on design, but place more emphasis on the knowledge and proper use of plant material. Graduates of these courses usually work for nurseries or retail garden centers that provide their customers with a landscape service.

However, if the individual is primarily interested in design, he should consider a school that offers a degree in landscape architecture. Usually the courses include some instruction in plant material and its use, but frequently the landscape architect relies on the nurserymen for information concerning specific plant material. This is not to imply that landscape architects are not familiar with plant material or that horticultural majors from state colleges can not provide an overall design. Both interest and on-the-job experience are as important as the individual's schooling to that individual's total knowledge.

Any individual considering either landscape design or architecture should take courses in mechanical drawing and surveying. Mechanical drawing will familiarize you with the various drawing tools and will increase your proficiency in both drafting and lettering. Such proficiency is necessary, since the completed plan that is presented to the client must be neat and present an attractive design with symbols and letters that are legible.

Surveying courses will familiarize you with the use of a transient, which is an instrument used to determine elevations. The existing and planned elevations of a proposed landscape are necessary to determine the proper grades that allow for water runoff from both land and paved areas, and the difference in elevation from one area to another may also be required to compute the amount of soil to be moved to raise or lower the eventual areas.

For convenience, a landscape plan can be divided into three areas: the public area, the service area, and the private area. The *public area* is that portion of the overall landscape between the street and the house. The *service area* may be beside the house or a portion of the landscape set aside either in the side or rear lawn area. The *private area* is the remaining portion of the landscape behind the house and may include areas beside the house.

During more formal periods of landscape planning, cone-shaped evergreens were positioned on either side of the front door and at the corners of the house, and formal hedges usually lined the driveway. Presently, with our more informal life styles, landscaping has become less exacting, and more imagination has been used to create the final plan to the point where the public area may not necessarily be a neatly planted front lawn and residence to be viewed from the street.

Depending upon the space available and the owner's wishes, the area between the house and street may be shielded from the street with formal or informal plantings, resulting in a private area for the owner's personal use.

Generally, if there is enough land available at the side or behind the house for the owner's needs and there is no necessity to shield the house from objectionable views or noise from the street, then the area is designed with the view from the street in mind. The view that the owner receives as he approaches the house and the view from within are also both considered in preparing the plan.

Both the service and private areas are designed with the owner's needs and wishes in mind. Prior to the advent of dryers, washers, and trash compacters, the need for a large screened service area was common. Now few people hang the wash out to dry, and even if the trash has not been compacted, the garbage cans are often confined to a small area either within or near the garage.

The vegetable garden area and compost pile are the more likely subjects of a screened service area. However, certain people derive great pleasure from their vegetable garden and prefer to view it from the patio or from within the house.

The private area has probably undergone the least change, mainly it should always be designed for the owner's use and pleasure. This use varies with individuals and certain cyclic trends, but certain items, such as outdoor eating and sitting areas, the children's play area, swimming pools, flower gardens or borders, and tennis courts, still retain popularity.

The amount of landscape planning required will depend upon several factors: the access from the house and possibly the parking area, the topography of the land, the number and types of constructional items to be included, existing and planned vistas, and the quantity of privacy desired.

After meeting with the clients and learning their desires and requirements, the planner prepares a plan to illustrate what will be done to their landscape. Often the client will have a plot plan, which properly positions the house on the lot plus a plan of the house. When both are available, measurements can be obtained directly from these plans, eliminating the need for actual measurements. If unavailable, then the actual measurements must be made and recorded.

Existing trees, rock outcrops, and property boundaries are also noted and eventually shown on the plan. When the measurements are used, scales of one inch equaling eight feet or one inch equaling 10 feet are most frequently used. A scale of one inch equals four feet allows for more detail, but it is only practical for a small area, since the resulting plan would be too large and cumbersome if the overall property were involved. For extremely large properties, a scale of one inch equals 20 or 40 feet may be necessary, with details of specific areas presented on additional pages with a scale employed that permits the necessary detail.

Once the house and other existing features have been drawn on the plan, the design is begun and the driveway, if not already in place in the landscape, will be one of the first features to be considered. The design of the driveway will depend upon the location of the garage, the distance from the street, and the size and shape of the property. If possible, it is advisable to provide a turn court so that automobiles can be backed into it, enabling them to always be driven into the street. Whenever the distance from the garage to the street is short, a turn court is simply not possible in the landscape plan.

Some owners will insist on a circular driveway even when the property is too small for it to be aesthetic. One definite advantage of the circular drive, however, is to properly direct visitors to the front door. Frequently, when the front walk is brought in from the driveway at some point near the garage, the walk may pass a side entrance, which may confuse the guests. To delineate the main entrance and separate it from a side entrance, the width of the walk may be enlarged or well lighted.

The constructional features in the public area are usually completed once the driveway, sidewalks, and lights have been properly positioned. The planting design will complete this area. Shade tree placement should be the first consideration, and unless there is a definite need for shade on the front of the house, the overall design is generally more attractive if the trees are placed at the sides of the house to frame it. Trees positioned in the center of the lawn or near the house are inclined to obstruct it, whereas trees placed near the street do not present the same problem, since the house is viewed beneath the branches and the relationship of the tree's size to the house is greatly reduced.

If the property is small, the side border plantings may be kept to a minimum to allow the adjoining lawn areas to meet and provide a more spacious effect. If, however, the adjoining properties are unattractive or the client prefers privacy, the side borders can be full and may extend across the front to screen the house from the street. Berms, or narrow ledges, can be added to reduce street noise and add interest to certain properties.

The front foundation planting will depend upon the design of the house and the owner's desires. Houses with a low silhouette and very little foundation exposed may require only a few shrubs planted in ground cover beds to unite the planting. Usually the shrubs are positioned on each side of the front entrance, at the corners and strategically along the foundation. The entire foundation may be covered if the foundation is apparent or if the client is a plant enthusiast, and can consist of one or two rows of plants.

The selection of a limited number of different plants to be used in the foundation planting is recommended. It is a definite mistake to plant the foundation area with a large variety of plants, suggesting an arboretum. Repetition of plants from the limited selection will aid in the creation of a harmonious design.

When the plants are spaced on the plan, the ultimate size and growth habit must be considered. It would be impractical to allow for maximum size in any foundation planting. Various forms of spreading yews, for example, are common in foundation plantings, yet even the dwarf yew will grow to a spread of 20 feet. It is best to plan for a spacing that will provide an attractive appearance when a particular area is completed. As the plants grow, they should be pruned to retain the desired height and shape.

The positioning of plants should adhere to the premise that the general growth habit should conform to the area involved; i.e., plants that are naturally upright should not be used beneath low windows, because too much maintenance will be required.

When upright, needled evergreens are desired and yews or arborvitae do not provide the desired effect, hemlocks are often used. When hemlocks are used in a foundation planting, they require yearly care, for they can grow to a height of 90 feet, necessitating their removal.

Courses on plant material provide the information on size and the conditions necessary for individual types of plants, and also list those that will withstand wind and excessive moisture or dryness. With experience, the designer learns which plants adapt best and which ones can be easily controlled.

The same design procedure is followed for the service and private areas of the property. The service area is usually small and can be unobtrusively worked into the design by screening with a planting or fencing or a combination of both.

The private area can be the most interesting part of the overall design, since there are several features that the client is likely to want included. These features should be placed to allow for convenient and logical traffic flow, yet arranged to permit maximum beauty and pleasure.

The outdoor sitting area is usually near and often attached to the rear of the residence. Terraces can be planned so that space is left for plantings near the residence or pockets are left within the terrace for various plantings.

Decks are normally attached to the house, and planter boxes are frequently employed to soften the architectural lines of the deck. They may also provide a sitting area with plant interest.

The views from the terrace or deck and from within the home are the most important, and other features should be positioned accordingly. It is equally important that a children's play area be viewed from the kitchen and from the outdoor sitting area, but it is not necessary to place this area in the most strategic viewing area. The primary view would include the rose or flower garden with its background plantings.

Some clients want the swimming pool as the main feature of their private area. This is not practical in the north, where the pool is drained or covered during the winter and should not be an item to feature.

The client's desires and budget will be the determining factors of the extent of the landscape plan involved for their property, and the landscape planner will be responsible for a functional design with aesthetic appeal.

After the rough design has been worked out, the landscape planner completes the plan and submits a finished copy to the client. The architectural features are labeled and are usually easy to identify, but often there is not sufficient space on the plan to write in the names of all the suggested plants and their quantities. Because of this, most designers use symbols for plants and keys to identify the type and quantity of plants used.

Each designer has favorite symbols that he is accustomed to using. Deciduous trees can range from a simple circle to circular forms with shading. Symbols for evergreen trees are generally circular in form with a spiky outline. Symbols for deciduous and evergreen shrubs are usually modified versions of the tree forms. Hedges can be represented by parallel lines with relatively plain lines being used for deciduous plants and spiky lines for evergreens. Ground cover areas may either be labeled or keyed to the plant list. Too much detailed drafting of the individual plants within the plan may prove confusing and may detract from the overall design.

It is important that each plant or group of plants and quantity be identified. One method is to assign each plant type a number. If the plant list is presented alphabetically, the first plant listed becomes number one, and whenever this plant appears on the plan, the number (1) is placed within the symbol or outside the symbol with a line drawn to the symbol.

If there is more than one plant in a group, e.g., three, the key number (1) would appear first with the quantity, or 3, second, for example, (1–3). This can be confusing if the numbers are reversed and incorrectly taken as one plant of item (3) from the plant list.

One method of avoiding possible confusion is to assign each plant a letter key. The first initials of the plant's Latin name are used for each plant. Thus, a Canadian hemlock (Tsuga canadensis) is keyed (TC). The symbol (TC) is used on the plan whenever the plant is suggested, and if there is more than one to be planted in a group, the key and quantity are then written (TC-3).

An example of a finished plan is presented in Fig. 14-1.

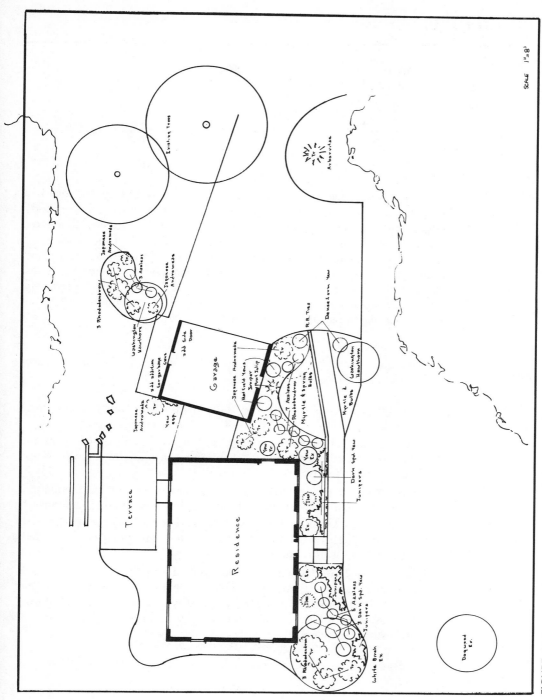

FIGURE 14-1
Completed landscape plan.

99

Chapter Fifteen

Lawns

The appearance of all residential properties is greatly enhanced by the presence of a smooth, green, well-manicured lawn. Flower beds, hedges, and foundation plantings are impressive only when complemented by a thick green carpet covering the ground on all sides of the house.

From the more than one thousand members of the grass family growing in this country, approximately 30 species have been selected for lawns. Which of these 30 species should be incorporated into a specific lawn can be determined only after an honest appraisal of the environmental area and its eventual maintenance treatment has been made.

GRASS SELECTION

Selection of which grasses to use with respect to the specific area involves consideration of the following:

1. *The climate*. Grasses with the ability to thrive in northern locations will be unable to survive in a hot, dry, southern location.
2. *Soil reaction*. Bent grass, redtop, and orchard grass desire a low pH (5.5 to 6.5), whereas bluegrass, the fescues, and clover prefer a pH that is only slightly acidic (6.5 to 7.0).
3. *Available light*. In shady areas, rough bluegrass and the fescues are sown, whereas in most sunny locations, bluegrass is the best grass.
4. *Soil moisture level*. Redtop, rough bluegrass, and bent grass can exist and actually thrive in poorly drained soils. Bluegrass, however, will grow only in well-drained soils.
5. *Usage*. For those lawn areas designed primarily for play, grasses such as Canada bluegrass and carpet grass are heavily blended with the bluegrass because of their ability to withstand heavy use and abuse. For lawns primarily for show, Kentucky bluegrass and its varieties are usually selected.

Selection of grass based upon the eventual maintenance treatment involves:

1. *Soil fertility*. Individuals who intend to follow the recommendations with respect to quantity and frequency of fertilization should have

lawns composed predominantly of bluegrass and bent grass. In contrast, lawns of low fertility levels should be either redtop or the fescues.

2. *Grass height.* Grasses that require close mowing [one inch (2.5 cm) or less in height] are bent grass, carpet grass, Zoysia grass, and rough bluegrass. Those that should be left at a minimum height of 1½ in. (3.75 cm) are the fescues, redtop, bluegrass, and Canada bluegrass.

Once the proposed lawn site has been evaluated with respect to the considerations just mentioned, the establishment of a *new lawn* can be undertaken. To accomplish this task, there are several requirements for success. For simplicity, they are presented in the order of their priorities.

1. Allow for proper drainage. Grass will not grow properly in waterlogged soil, since its roots need air as well as moisture. Low areas in the soil must be built up to prevent water from lying on the soil and destroying the grass. Existing trees that lie below the eventual grade can be saved by placing a protective well around the trunk. To prepare this well, large stones are placed on the original soil layer. Smaller stones, followed by cinders, are placed on top of the large stones and are then covered with soil to make the final grade. To insure drainage, tile drains are positioned so that they radiate out from the well (Fig. 15-1).

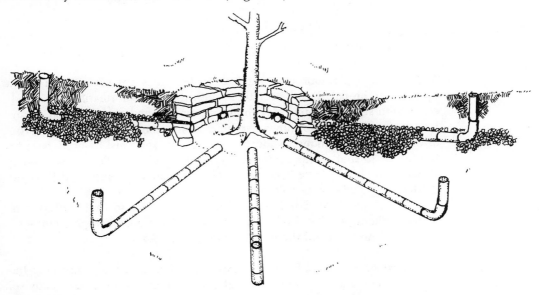

FIGURE 15-1
Treatment of a tree below the final lawn grade.

Tile drains may be needed to remove excess water from the eaves of the house, so that the land surface is not eroded away during periods of heavy rainfall. Plastic perforated pipe is laid down below the soil surface. It is connected to the end of the downspout at the corner of the house and led away from the house on a downward slope at least one 8-ft (2.4-m) length of pipe (Fig. 15-2).

FIGURE 15-2
Positioning of tile drains.

2. Establish a pleasant and practical grade. It is a misconception that a lawn has to be flat. It is better to attempt to establish a harmonious landscape with sloping contours, both for appearance and proper runoff of surface water. Figure 15-3 shows two houses, one above street level and the other below. In the house below street level, the soil must be graded up to the house so that water drains away from the house. The house above street level should have a nearly level area near the house and then be graded on a gentle slope to the street.

All slopes should not be graded too severely, so that the lawn is scalped when cut (Fig. 15-4) but should receive a gentle sloping curve so that the mower cuts all the grass evenly (Fig. 15-5).

3. Additions to the soil. If the existing soil is either a dense clay that exhibits poor drainage when wet and cracks when dry, or a very sandy soil unable to hold moisture, topsoil should be added. Frequently, development houses have had their topsoil completely removed, an undesirable subsoil being left on the surface. It is usually necessary to cover this subsoil with topsoil. The addition of any organic matter will be beneficial, since soils used to support the growth of grass should be at least ten percent organic.

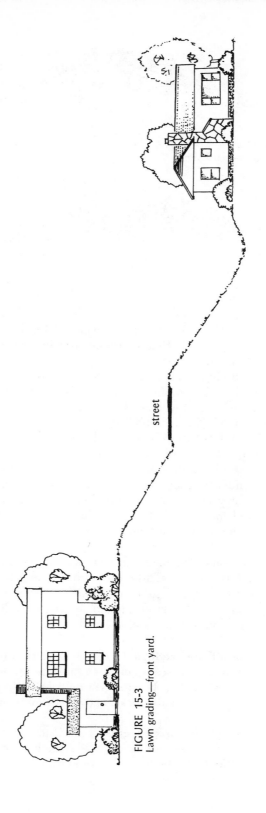

FIGURE 15-3
Lawn grading—front yard.

street

FIGURE 15-4
Lawn grading—too severe.

FIGURE 15-5
Ideal lawn grade.

A soil acidity test should be made, and if the soil has a pH below 5.5, lime must be added. Finely ground, pulverized limestone is the best for use on lawn areas.

A complete fertilizer, containing nitrogen, phosphorus, and potassium, and having a minimum composition of 10-6-4, should be added to all lawn soils. One pound (0.45 kg) of nitrogen is the recommended amount that should be added to 1000 ft² (92 m²) of lawn surface. If a 10-6-4 fertilizer is used, 10 lb (4.5 kg) will be necessary to satisfy this area. Lawn fertilizer is sold in a variety of compositions. Some are totally organic and are slow to release their nutrients to the lawn (these are usually temperature dependent), whereas others are chemical, inorganic ones that are extremely water soluble and supply their nutrients rapidly. A blend of both organic and inorganic materials in the same fertilizer is now generally accepted as the most ideal type to add.

4. Preparation of the seedbed.   The soil and its additives (organic matter, lime, and fertilizer) should be turned over and well mixed. This procedure is accomplished by the use of a rototiller, a power cultivator which ensures that the top 8 in. (20 cm) of soil and additives will be well mixed. A garden spade may be used if the area is small.

5. Leveling off the seedbed.   Once all the additives and soil have been thoroughly mixed, the complete area must be leveled off by raking or rolling the area. Watering the area will help to settle the soil and is strongly recommended. Dirt clods, rocks, and weeds must be removed at this time.

6. Selection of the proper grass.   For sunny exposures, it is strongly suggested that a grass seed mixture be selected that contains a minimum of 45 percent Kentucky bluegrass. It is not advisable to purchase a "cheap" mixture, especially when a lawn is considered a long-term investment. Most grass mixtures contain at least one premium grass, such as Kentucky bluegrass, that may not germinate for six weeks and one or more grass types, such as a ryegrass or fescue, selected because of their rapid germination rate. These types are added to ensure a quick cover for the soil surface before the premium grass appears.

For shaded lawn areas, the fescues are the best to use. It is not wise to select an inexpensive mixture.

To properly seed a 1000-ft^2 (92-m^2) area, the following amounts are usually recommended: 4 lb (1.8 kg) of Kentucky bluegrass, either alone or in a mixture, 3–6 lb (1.35–2.7 kg) of ryegrass or fescues, and 2 lb (0.9 kg) of Merion bluegrass or Winsor.

7. Time of planting. The most favorable time to seed a new lawn is early fall. The new seedlings have fewer weeds to compete with. It is possible to sow in the spring, but much more time and energy are required.

8. Seeding the lawn. The best time to seed a lawn is in the afternoon of a calm day. The seed should be divided into two equal halves. Either by spreader or by hand, sow one-half in a north-south direction, and the remaining half in an east-west direction. On steep banks and other areas susceptible to erosion, it may be more practical to lay sod than to sow grass seed.

9. After seeding is finished. The seed must be raked into the soil. Once this is completed, roll the entire area with an empty roller. The area seeded should be watered within 24 hours after sowing. The first few weeks after seeding are the most critical with respect to water. The seeded area should never be allowed to dry out.

To prevent the loss of newly germinated grass to bad weather conditions, such as heavy rains and extended dry spells, it is advisable to cover the seeded soil surface with either salt hay or burlap. Both conserve water and protect the surface against being washed away.

10. Mowing. The grass seedlings should be allowed to reach a height of 2½–3 in. (6.25–7.5 cm) before the first cutting. The blade of the mower must be sharp and the cutting height set at 2 in. (5 cm), although it can be set at 1½ in. (3.75 cm) if the grass mixture is mainly Merion bluegrass. The grass clippings should be removed from the lawn surface, since their presence may retard or suppress the growth of the new seedlings. Mowing should always be done when the lawn surface is completely dry. A reel-type mower is most suited to be used on a smooth, flat lawn surface, whereas a rotary-type mower is used when the terrain is uneven.

To maintain an *established lawn*, the following items are suggested for consideration:

ESTABLISHED LAWNS

1. Lawns must be frequently fertilized (in early spring, in June, and again in the early fall).

2. In those geographical areas subjected to freezing conditions, one light rolling, with the roller only half filled, may be required to press the grass crowns back into the soil.

3. Grass clippings must be raked up and removed, if the grass was too high when cut.

4. Aeration is the only way to cultivate the lawn. Special machines can be rented to accomplish this task. Aeration must be done before any crabgrass killer is applied.

5. Broad-leaf weeds are easily controlled by spraying with 2,4-D on warm afternoons. The air must be still when 2,4-D is used, since it may drift and harm other plants if applied when windy. The applicator selected must not be used for any other purpose and should be labeled accordingly. It must be mixed according to the manufacturer's specifications.

6. The best and most effective method of destroying crabgrass is the careful use of a preemergence crabgrass killer, a chemical additive that forms a protective coating on the soil's surface which prevents the germination of new crabgrass seedlings. When the lawn must be reseeded, the chemical siduron (Tupersan) is applied in May or early June after seeding has been done. On established lawns where no seeding will be done, bensulide (Betasan or Presan) is applied at any time from March through May.

7. Lawns should receive a good *soaking* at least once a week. Rainfall may be sufficient to satisfy this requirement, but during periods of drought, the lawn must be sprinkled.

8. An application of diazinon destroys all the lawn insects that are capable of surviving throughout the winter in the soil.

9. Proper cultural practices are the best protection against turf grass diseases. The following steps are helpful in avoiding turf disease problems without resorting to expensive fungicides.

    a. Plant naturally resistant varieties, such as Kentucky bluegrass, red fescue, and perennial ryegrass.

    b. Avoid excessive applications of nitrogen, particularly in the spring.

    c. Use a leaf catcher on your lawn mower to help reduce thatch buildup.

    d. Periodically remove thatch buildup to restore vigor and to remove decaying organic matter.

The chemical control of turf disease is expensive and time consuming, because the chemicals give only temporary relief and must be applied repeatedly. The identification, classification, and treatment of any turf disease represents a problem for the homeowner or landscaper untrained in plant pathology. The following turf grass diseases are the most common ones:

a. Helminthosporium leaf spot and crown rot.

b. Fusarium patch or snow mold.

c. Sclerotinia dollar spot, copper spot, or red thread.

d. Rust.

e. Powdery mildew.

10. Correct mowing procedures are an asset to any fine lawn. Close mowing reduces root growth and increases the possibility of weed infestation. Mow lawns at intervals, depending on the speed of growth, and be certain that the lawn mower is sharp and set at the proper height, because most lawns, with the exception of Kentucky bluegrass and bentgrass putting greens, should not be cut closer than 1½ in. (3.75 cm). It has been found that cutting half or more of the foliage of grass causes the root growth to stop for a time, and this stoppage varies in proportion to the percentage of foliage removed. Of the two common types of lawn mowers, rotary and reel, most turf experts prefer a sharp, well-adjusted reel type for giving a lawn a well-manicured appearance. Rotary types are best when the grass is extremely high, since a reel type tends to clog under these conditions.

11. Renovate and repair all lawn areas that have been damaged. This practice usually entails repairing the damage that has been done each year. Injured areas can be nursed back to health without jeopardizing any healthy turf in adjacent areas. Whether the damage has been caused by starvation, rough usage, serious drought, disease, or insect infestation, the following information will usually supply the remedy:

a. Rake out all the dead grass and break up the surface thoroughly, even though some good grass plants are dislodged.

b. Reestablish the existing contour of the lawn surface by careful raking.

c. Add humus where needed, and mix it thoroughly with the topsoil.

d. Sow the grass seed uniformly over the entire area, following the rate listed on the package.

e. Apply lime and fertilizer at the recommended rate.

f. Rake and roll the area lightly to imbed the seed and follow through by watering and giving the newly seeded area the same care as previously mentioned for new lawns.

Problems can result in a lawn area, even though all the necessary directions and requirements have been obeyed. In the following paragraphs, specific lawn problems and their solutions are discussed.

Acidity. A soil pH below 5.5 prohibits proper growth of grass. An application of ground limestone usually corrects the problem.

Chinch bugs. Their initial infestation is usually observed as circular dead areas in the lawn. To confirm their presence, the soil area must be turned over and examined. To correct this problem, the soil should be drenched with diazinon or a related product.

Dog manure burn. This is usually a temporary injury that most grass varieties can overcome. However, when lawn areas are continually overrun by dogs, a more serious problem can result.

Fertilizer burn. This condition is caused by too heavy an application of fertilizer. Provided the grass roots are not injured seriously, the lawn should bounce back.

Fertilizer skipping. This condition usually does not become apparent until long after the fertilizer has been applied. It is usually caused by some malfunction of the applicator. The area missed should receive a fertilizer application as soon as possible.

Moles. Their presence in the lawn is usually due to an abundance of grubs and insects living in the soil. Removing their source of food usually results in their disappearance. An application of diazinon should hasten their departure.

Moss. If your lawn contains moss, its presence can be due to too acid a soil, an infertile soil, or a damp, heavily shaded area. In certain cases, all three factors are interacting to provide the proper environment for the growth of moss. The addition of both lime and fertilizer should correct most problems related to moss. In areas of dense shade, it is advisable to prune the surrounding plant material, so that more sunlight can fall on the lawn surface.

Mushrooms. Their presence indicates that there is decaying organic matter buried beneath the soil which supports the growth of fungi organisms during periods of ample water and high humidity. The simplest solution is to dig up the buried organic material and dispose of it.

Poison ivy. This persistent weed can give grief to many homeowners. Repeated applications of amino triazol have been found to yield the best eradication.

Thatch. This buildup of previously cut grass parts can seriously deter the normal growth of grass. When it becomes heavy on the lawn surface, it should be removed. To minimize its buildup, the grass should be cut when dry, at a height one-half inch above the mower setting, and collected in a leaf catcher.

Chapter Sixteen

Specimen Trees

Trees are planted for a variety of reasons. They are extremely important in complementing the landscape design of the property. Some trees are grown for their beauty, for their form, or for their foliage or flowers, whereas others are planted to provide shelter from the wind or shade from the sun.

TREE SELECTION

In selecting the ideal tree for a specific place, there are many points to consider before arriving at a decision. The size that the plant will reach at maturity, its rate of growth, its adaptability and hardiness to a particular climate, and any undesirable traits displayed by the tree during its life are all important. There are trees that flower, those that provide shade, and certain ones noted for their production of colorful fruits or edible nuts.

The final selection must be based upon a blending of the wishes of the homeowner, his pocketbook, the desired location, and the plants available. It must be remembered that a properly planted tree increases in value with age. The exception would be when an improperly selected tree matures and completely overwhelms its surroundings.

The considerations expressed above are equally important when one is selecting the proper shrubs needed to complete the total landscape picture.

The first type of specimen tree to be considered in this presentation is the shade tree. Most shade trees require a spread of 50 ft (15 m) and should be planted 30 ft (9 m) from the house. If possible, it is advisable to plant this type on the south or west side of the house to ensure its effectiveness as a shade tree. A shade tree should frame the house and serve as a background for it, but should never be positioned so as to completely hide the house (Fig. 16-1). Deciduous trees should be positioned so that their branches will not overhang the roof and fill the gutters with falling leaves. It has been experimentally proven that shade trees can lower indoor temperatures by 20°F (11°C) on hot sunny days.

Using the common names as the reference, the following is an alphabetical presentation of the shade trees grown in temperate areas. (Consult the appendix for additional listings of trees commonly grown in other geographical locations.)

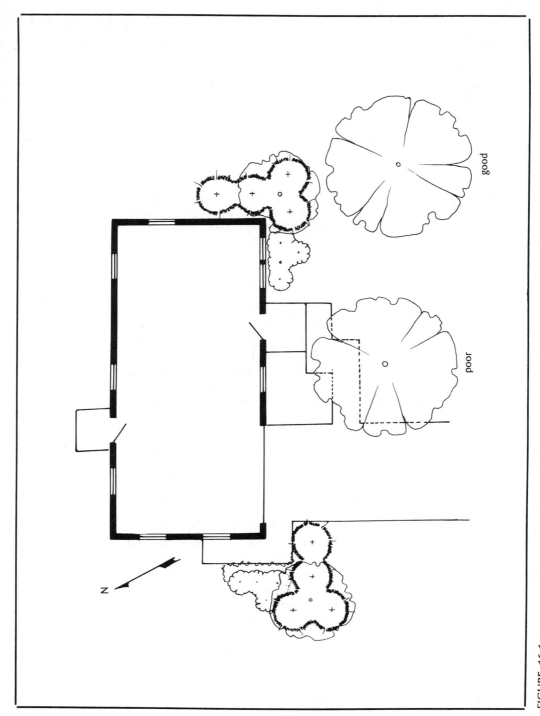

FIGURE 16-1
Position of shade tree in the landscape.

110

American linden or *basswood* is a rapid grower with small yellow flowers. Its eventual height is 130 ft (40 m). Its growth may suppress the growth of grass underneath the tree. This tree requires abundant moisture and a deep, fertile soil and is able to tolerate temperature extremes. *Tilia americana* is its scientific name.

Crimson king maple or *Acer platanoides variety Schwedleri* is a more desirable variety of Norway maple, because of its dark red leaves. At maturity, it may reach a height of 90 ft (27.5 m).

European beech or *Fagus sylvatica* is noted for its leaves, which remain on the tree for most of the winter. In the fall, the leaves first turn yellow and then change to brown. Other notable features are its gray-colored bark, its thick trunk, and lower branches that touch the ground.

Maidenhair tree or *Ginkgo biloba* has fan-shaped leaves that turn yellow in the fall. An unusual feature of this tree is that it has both male and female plants. The male form is the one that should be selected, since the odor of the female fruits is unpleasant. It is a slow-growing tree.

Honey locust or *Gleditsia triacanthos* is a picturesque tree with thorny stems characterized by rapid growth and yielding adequate shade. The *moraine locust* is a more recent variety that is now preferred because it lacks thorns, allows for the growth of grass beneath it, and is free from disease. Its growth in height averages about 3 ft (0.92 m) per year, and it reaches a maximum height of 100 ft (31 m). Another desirable feature of this tree is that its tiny leaflets dry up and disintegrate into the lawn upon falling in early autumn, thus eliminating the tedious fall leaf-raking chore.

Japanese maple or *Acer japonicum* is another small shade tree, reaching a mature size of 25 ft (7.6 m). The leaves may be green, purple, or red in color, and some varieties have finely divided leaves.

Katsura-tree or *Cercidiphyllum japonicum*, a native of Japan, is grown for its foliage. Its delicate, heart-shaped leaves first turn a bright yellow and red in the fall. It is characterized by having a broad dome and branches that touch the ground.

Norway maple or *Acer platanoides* is a rapidly growing tree that provides very dense shade. Its leaves are green in color, changing to shades of yellow during the fall.

Pin oak or *Quercus palustris* is the most popular oak because it is easy to transplant, has a symmetrical pyramidal form, and grows rapidly. Its leaves may show chlorosis, which can easily be corrected by a soil application of chelated iron. At maturity, its height should approach 80 ft (24.5 m).

Sugar maple or *Acer saccharum* is noted for its dense shade, brilliant fall color, maple syrup and sugar, and a steady growth rate, reaching a mature height of 100 ft (31 m).

Sweet gum or *Liquidambar styraciflua* is a reasonably fast-growing shade tree offering brilliant fall color, corky bark, and a pyramidal shape. It grows to a height of 80 ft (24.5 m) at maturity.

Sycamore or *Platanus occidentalis* is identified by its two-toned bark, large leaves, and erect growth. One of its hybrids, *London plane tree or*

Platanus acerifolia, is a very dependable tree able to survive in New York City. An objectionable feature for some is the clutter resulting from its shedding bark and falling fruit heads. It is able to thrive in a variety of soils and when mature forms a broad, desirable shade tree, reaching 100 ft (31 m).

The second type of specimen tree discussed will be the flowering tree. As a general rule, this type of tree is usually planted at the corner of the house, near a patio, or in a large spacious front yard. The beauty of flowering trees is best enhanced by positioning them to blend with other flowering shrubs and evergreens (Fig. 16-2). The minimum distance between two flowering trees should be 12 ft (3.7 m).

Catalpa speciosa is more noted for its cigar-shaped fruit than for its flowers. It is city hardy and may reach 50 ft (15 m) in height. This tree prefers a well-drained, moist, fertile soil but can tolerate drought and alkali soils.

Flowering cherry trees, genus *Prunus*, offer the homeowner a good selection. Flowers may be single or double, pink or white in color, while the trees may vary from an upright habit to a weeping one. Cherry trees

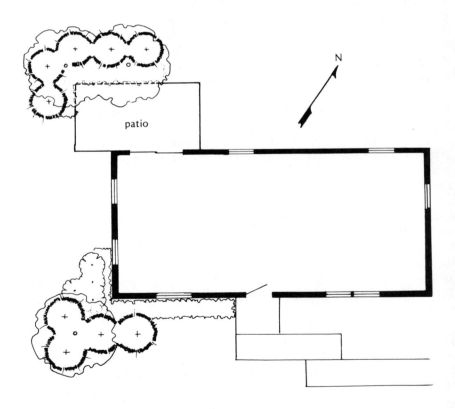

FIGURE 16-2
Positioning flowering trees in the landscape.

are very tolerant of temperature extremes and grow best in rich, well-drained soils and full sun.

Crabapples, genus *Malus*, display small apples after flowering has passed. Pink or red flowers appear with the leaves in April. Crabapples are able to grow and thrive in a wide range of soils and are not affected by fluctuations in temperature.

Dogwood or Cornus florida may have white, pink, or red flowers in the early spring, giving way to attractive foliage on horizontal branches throughout the summer months. As fall approaches, the leaves change colors and red berries appear on the stems. They are tolerant of a wide variety of soils, provided the soils are moist and well drained. Dogwoods should be transplanted only in the spring of the year.

European mountain ash or Sorbus aucuparia is a small tree whose clusters of white flowers give rise to its characteristic orange fruit in the fall. It is a slow-growing tree with compound leaves that may reach a final height of 40 ft (12.2 m).

Golden chain or Laburnum vossi is noted for its yellow chain-like flowers appearing in May. It is also chosen for a site requiring a mature height below 25 ft (7.6 m).

Goldenrain tree or Koelreuteria paniculata has clusters of yellow flowers during June and July, making a pleasant contrast with its light green compound leaves on twisted branches. This tree needs full sun exposure and is pest-free and drought resistant. It reaches a mature height of 20 ft (6.1 m) and a spread of 15 ft (4.6 m).

Horse chestnut or Aesculus hippocastanum is desirable for its dense shade, its white flowers tinged with red appearing in May, and its characteristic nut in the fall. It makes an excellent lawn specimen and street tree because of its uniform size and shape. It prefers plenty of moisture, a sunny exposure, and some protection from the wind. For some, it may be undesirable because of the litter caused by the dropping of its leaves and fruits. Its eventual height will be near 80 ft (24.5 m).

Japanese tree lilac or Syringa japonica is a small, compact tree that bears large trusses of white flowers in June. It may reach a mature height of 25 ft (7.6 m). It is free from injury due to pests and requires little maintenance.

Mimosa or Albizzia julibrissin is a small, graceful tree with fernlike leaves and pink flowers forming in early summer, eventually changing into its characteristic podlike fruit. It will not survive under extremely cold winter conditions. It is a fast grower, reaching a flat, top height of 25 ft, (7.6 m), making it an ideal patio specimen.

Flowering peach and plum trees both belong to the genus *Prunus*. Their desirability lies in their showy flowers, their rapid growth, and a mature height not exceeding 10 ft (3 m), making them ideal for small areas.

Redbud or Cercis canadensis is one of the first trees to flower in the spring. Its deep pink flowers appear before the foliage comes out. Because its growth habit is irregular and its eventual height may not exceed 25 ft

(7.6 m), it either can be used as a specimen tree or can be blended among a shrub planting.

Saucer magnolia or Magnolia soulangiana yields lavender blossoms in May that cover the entire tree before the leaves appear. Being relatively disease-free, they do well in a city environment.

Silverbell tree or Halesia carolina is a small, rapidly growing tree that produces a multitude of small, white, bell-shaped flowers in late spring. The flowers give rise to fruits that are both interesting and desirable to squirrels. It is free from disease and has a shrublike habit of growth, reaching a mature height of 30 ft (9 m).

Star magnolia or Magnolia stellata is a small, slow-growing tree that produces fragrant white blossoms early in the spring that resemble a star. Its height usually does not exceed 15 ft (4.6 m).

Scarlet hawthorn or Crataegus oxacantha and Washington hawthorn or Crataegus phaenopyrum are characterized by horizontal branches with thorns, a profusion of spring flowers, either red or white, applelike fruits that appear in the fall, and brilliant fall color. These trees can be used as specimens, in an untrimmed hedge, or as a windbreak.

Smoke tree or Cotinus coggygria has purple-pink filmy flowers appearing in July and lasting throughout the summer. As they dry, they turn brown, giving a smokelike effect. Because of its shrublike growth habit, reaching a maximum height of 15 ft (4.6 m), it can be used as a tall shrub or small tree.

Sorrel tree, sourwood or Oxydendrum arboreum is a small, slow-growing tree producing strings of white bell-like flowers, similar to Pieris. It is usually selected for its brilliant crimson fall color.

Sourgum or Nyssa sylvatica also has brilliant crimson autumn foliage. Its flowers, appearing in May, are relatively inconspicuous. This tree may grow to a mature height of 100 ft (31 m).

Tulip tree or Liriodendron tulipifera receives its name from the greenish-yellow tulip-shaped flowers that appear during the summer months once the tree is at least 10 years old. It must have plenty of moisture and deep soil for its root system. It is characterized by growing tall and straight, reaching a mature height of 70 ft (21.1 m). Because it is tall, grass is able to flourish beneath the tree.

Wisteria tree or Wisteria sinensis is considered more a vine than a tree, even though it can grow rapidly to 20 ft (6.1 m). Its large purple or white pendulous flowers appear in profusion in June and give rise to its characteristic pod-shaped fruits.

The third category discussed will be *evergreen trees*. They are chosen for the following reasons:

1. They make ideal backgrounds.
2. They offer privacy.
3. They are used as windbreaks.
4. They can hide ugly views.
5. They provide a variety of colors and shapes to the landscape.

Alberta spruce or Picea albertiana conica is a small, close-needled conical tree that can be used in foundation plantings or as a specimen tree.

Austrian pine or Pinus nigra can be used as a specimen, a windbreak, or as a shade tree, because it can grow to a height of 100 ft (31 m). It is hardy, free from disease and pests, and is durable.

Canadian hemlock or Tsuga canadensis can be planted alone, in a hedge row, or as a clump. It is important to give it plenty of growing room so that its grace and beauty can be appreciated. It responds well to shearing and is free of insects and diseases.

Colorado spruce or Picea pungens is normally grown as a specimen tree but may be planted as a windbreak. Its popularity is due to its brilliant blue needle color. When mature, it reaches a height of 100 ft (31 m).

Douglas fir or Pseudotsuga taxifolia is sometimes preferred as a specimen tree to spruces, because of its graceful habit and its ability to withstand shearing to make it more dense and compact. If left untouched, its height could eventually exceed 300 ft (92 m).

Japanese black pine or Pinus thunbergi is selected for its unusual growth habit, as well as for its ability to withstand wind. It can be used as a specimen tree or can be planted as a windbreak.

Pfitzer's juniper or Juniperus chinensis Pfitzeriana will usually not exceed 10 ft (3 m) in height, but may spread to 20 ft (6.1 m) in width. Other desirable features are its blue-green color and its adaptability to shearing into a variety of shapes.

White spruce or Picea alba may be used as a specimen tree or as a windbreak. It has a dense, pryamidal growth habit, reaching a mature height of 100 ft (31 m).

Pendulous or weeping trees comprise another group of specimen trees.

Weeping birch or Betula pendula cutleaf offers a tree with white bark and deeply cut leaves, in addition to its weeping nature. It is ideally suited for shade or show.

Weeping beech or Fagus sylvatica borneyensis has all the desirable features associated with the European beech plus the added one of pendulous branches. The flowers of this tree are formed on separate male and female trees. This plant forms long tap roots, which makes it almost impossible to be moved once it is planted.

Weeping cherry or Prunus subhirtella is selected for its mass of pink spring flowers as well as for its weeping habit. The weeping part of the plant is grafted onto a tall straight trunk of another hardy cherry variety. It grows rapidly to an eventual height of 30 ft (9.2 m).

Weeping willow or Salix babylonica is a tree that thrives in wet places, but can grow in almost any climate and is able to tolerate extremes in temperature. It is the most popular pendulous tree grown, even though its branches and leaves clutter the ground after being exposed to moderately heavy winds. When mature, it reaches a height of 40 ft (12.2 m) and a spread approaching 70 ft (21.1 m). Because of this fact, it is necessary to provide this tree with ample growing and viewing room.

European white birch or Betula pendula is noted for its beautiful trunk. It is a fast-growing tree with showy fall color. This tree grows best in a well-watered location, but is able to adapt to a wide range of climatic conditions.

Fall color is due to the combination of lower fall temperatures and the formation of an abscission or separation layer at the base of each leaf. Reduction in day length is responsible for the formation of this abscission layer, which restricts the flow of materials into and out of the leaf. Once this occurs, photosynthesis is retarded, causing a fading of the chlorophyll which permits the chemicals left trapped in the leaf to react with the available light, eventually producing the characteristic fall colors.

Hedges are designed to provide the home landscape with protection and shelter from the wind plus the added benefits of subtly entwining the components of the landscape into one permanent harmonious picture. Any plant material used as hedge material must have thick foliage with branches close to the ground and must have the ability to respond properly to shearing.

Some hedges are also planted to act as privacy screens and should be planted so that they do not detract from the natural beauty of the landscape. In recent years, plant screens have been replaced by walls constructed of wood or other building materials.

In the selection of hedge material, the gardener must consider whether or not the hedge should be evergreen or deciduous, and whether or not the hedge should be dense and impenetrable or one having space between the individual plants. These two desires are both related to the cost per eight-foot (2.4-m) section and must be handled as a long-term investment by the purchaser.

The following presentation is divided into two major categories: deciduous plants and evergreen plants. Information concerning the eventual height of the plant at maturity, and the correct spacing between plants will be covered for each plant. Other pertinent information that the purchaser should be aware of will also be presented.

The first deciduous material is *barberry or Berberis thunbergi*. It has become a perennial favorite for those individuals who desire a relatively low, dense, impenetrable hedge. When planted, the distance between the plants should not exceed 2 ft (0.6 m). It is characterized by small red berries and thorns. The leaves of these plants may be either red or green.

Burning bush, winged euonymus, or Euonymus alatus is noted for its brilliant red fall color and corky stems. When planted to form a hedge, the plants should be 3 ft (0.92 m) apart. They will reach a mature height of about 5 ft (1.5 m).

Another deciduous hedge material noted for its red berries in the fall is *cotoneaster*. Its planting distance and eventual height are similar to the

DECIDUOUS HEDGES

117

barberries. However, the growth habit of most cotoneasters is of a sprawling, random nature, whereas the barberry is primarily upright.

Forsythia is desirable for its profusion of yellow flowers in early spring. There are several varieties available, each with a slightly different growth habit. All the varieties selected as hedge material should be planted at 3-ft (0.92-m) intervals with a maximum height expected of about 6 ft (1.84 m).

Those who desire a hedge with a definite upright habit that can be sheared and kept at any height up to 15 ft (4.6 m) tall may select a variety of *privet or Ligustrum*. The distance between plants should not exceed 1½ ft (0.46 m). Because of this requirement, some gardeners prefer to plant privet in trenches rather than to dig individual holes.

A *rose* hedge is desirable for its continual display of color throughout the summer as well as for its impenetrability. Individual plants should be spaced at intervals of 3 ft (0.92 m), yielding an eventual height of about 6 ft (1.84 m).

A hedge that offers an upright habit with gray-green foliage is *Russian olive or Elaeagnus angustifolia*. These plants should not be used by those seeking a dense, low-growing hedge, since its mature height may exceed 20 ft (6 m). Most prefer to allow these plants their natural growth, and, therefore, refrain from trimming or shaping them in any way.

Several species of *spirea* are used as hedge material. All reach an eventual height close to 8 ft (2.4 m), and should be planted at a distance of 3 ft (0.92 m) between plants. Because of the drooping, random growth habit of the branches, only the top is pruned to achieve the desired height.

EVERGREEN HEDGES

The first evergreen plant used for hedge material is *arborvitae or Thuya occidentalis*. These plants, also known as white cedar, may reach a mature height of 30 ft (9.2 m). When planted, they should be placed 4 ft (1.2 m) apart. As a hedge, some of the plants may not grow at the same rate as the rest, leaving definite variations in their height and appearance.

Evergreen azaleas, when planted 3 ft (0.92 m) apart, will provide an interesting, low-growing hedge. They remain green throughout the year with the added advantage of being in color during the early spring. To insure the presence of flowers, any pruning that is required to maintain a certain desired height should be done immediately after flowering has finished.

Boxwood or Buxus has a similar growth habit to barberry, but has the added advantage of being evergreen and thornless. The Korean variety is more winter hardy than some of the other varieties that necessitate protection from severe winter conditions. These plants will reach a height of 8 ft (2.4 m).

When planted at intervals of 4 ft (1.2 m), *hemlock or Tsuga canadensis* matures into a dense evergreen hedge that can be maintained at any height. It can be sheared to give a well-manicured look, or it can be left natural with only the top clipped. It has no equal when an evergreen hedge taller than 10 ft (3 m) is desired.

Japanese holly or Ilex crenata is a broad-leaved evergreen that forms a dense hedge attaining a maximum height of 8 ft (2.4 m). The individual plants should be spaced at intervals of 2 ft (0.6 m) when planted.

When mature, *Pfitzer's juniper or Juniperus chinensis Pfitzeriana* forms a dense, impenetrable evergreen hedge. The distance between individual plants should not exceed 4 ft (1.2 m). Most juniper hedges never grow above 6 ft (1.84 m) in height.

Firethorn or Pyracantha coccinea is preferred by some for its orange-colored fruit that appears during the fall. The individual plants, spaced at 4-ft (1.2-m) intervals, need plenty of growing room and thus should not be used if a narrow hedge is desired. Its growth is easily controlled by pruning, but if left alone, it will reach a height of 10 ft (3 m). It is susceptible to scale insects, and must be sprayed frequently to control them.

When spaced at 5-ft (1.5-m) intervals, *Scotch pine or Pinus sylvestris* makes an excellent windbreak or screen. The individual plants are usually left untouched and will eventually reach a mature height of about 20 ft (6 m). As specimens they will reach a height of 80 ft (24.4 m).

Another excellent evergreen used as a windbreak or screen is *Colorado blue spruce or Picea pungens,* well known for its blue-green needles. Its eventual height can be kept at 6 ft (1.84 m) when trimmed, and when not trimmed, it could reach a height anywhere between 30 and 100 ft (9.2 to 31 m) depending upon the distance between the individual trees.

Several varieties of *Japanese yew or Taxus* are used for hedges. Some have primarily a narrow upright habit, whereas others produce a hedge that is wide, dense, and low-growing. It is selected by many because its dark green foliage provides an excellent contrast or background for flower gardens.

Chapter Eighteen

Ground Covers and Mulches

Many property sites contain areas where the seeding of grass is impractical, such as areas of deep shade, steep banks, rocky terrain, and areas where the roots of large trees use up the available moisture. In these areas, it is definitely beneficial to cover the soil surface with some kind of low vegetation.

The benefits derived from the planting of a ground cover are suppression of weed growth, assistance to erosion control, protection to existing plants against wide temperature fluctuations, and beautification of the area by a proper blending of plants having colorful blooms with those having deep green foliage.

When one is selecting a suitable ground cover, several considerations should be kept in mind. It should be inexpensive, perennial, easy to plant and maintain, highly adaptable to the prevailing climate and soil conditions, easily propagated, and vigorous enough to sustain itself and increase in number.

DESIRABLE GROUND COVERS

Carpet bugle or Ajuga reptans is a deciduous ground cover that is planted 9–12 in. (22.5–30 cm) apart. It reaches a height of 2–3 in. (5–7.5 cm), and is propagated by division. It spreads rapidly, yielding blue flowers, and has either green or bronze foliage.

Rockspray or Cotoneaster horizontalis is another deciduous plant with small leaves and bright red berries that last until the first frost. These plants are spaced at intervals of 1½–2 ft (46–61 cm), reaching a mature height of 2 ft (61 cm). Propagation is by stem cuttings.

Wintercreeper or Euonymus radicans reaches 9–12 in. (22.5–30 cm) in height but can be sheared lower if desired. It is a fast spreader, is evergreen, is propagated by stem cuttings, and is planted 2 ft (61 cm) apart.

A desirable evergreen ground cover reaching a mature height of 6 in. (15 cm) is *English ivy or Hedera helix*. It spreads fast and will climb if given support. The individual plants are spaced at intervals of 1½–2 ft (46–61 cm) when planted and are readily propagated from stem cuttings. In certain locations, it may be subject to winter burn.

120

Juniperus horizontalis or prostrate juniper is a low-growing evergreen having either green or blue-green foliage and never reaching a height greater than 12 in. (30 cm). Individual plants are spaced at 2–3½-ft (60–105-cm) intervals because of the rapidity with which it spreads.

A deciduous shrub yielding a variety of flower colors is *trailing lantana or Lantana callowiana*. New plants are obtained from stem cuttings. Their random growth habit and planting distance is similar to cotoneaster, but it reaches a height of 3 ft (91 cm) at maturity.

Japanese spurge or Pachysandra terminalis survives in a wide range of differing soils. It can be propagated by division, and by root or stem cuttings. It reaches a maximum height of 6 in. (15 cm), but is very slow to spread to open areas. Its leaves are evergreen and have an interesting desirable shape. Planting distance between plants should be 6–7 in. (15–17.5 cm).

A low-growing evergreen ground cover that forms a dense mat with showy flowers in the spring is *moss pink or Phlox sublata*. New plants are easily obtained by division. When planted at intervals of 1½ ft (46 cm), these plants will eventually spread to fill the open space.

Sedum acre or stone crop is a fast-growing, low-growing succulent that is easily propagated by division of existing plants and from cuttings. It is adaptable to a variety of growing areas, eventually forming a dense mat. There are numerous varieties grown, each yielding its own characteristic color.

A perennial herb that is frequently chosen as a ground cover is *lemon thyme or Thymus citriodorus*. It has a purple flower and a leaf that has a pleasant odor when either stepped on or pressed between the hands; it grows well in any soil. New plants should be planted at a distance of 6 in. (15 cm) apart.

Another low-growing evergreen is *myrtle, periwinkle, or Vinca minor*. It is chosen because it produces tiny blue or white flowers, depending upon the variety selected, because it spreads rapidly, and because it is able to grow well in poor soil. Individual plants are obtained by division or from stem cuttings and are planted at intervals of 12 in. (30 cm).

Presented in Table 18-1 are the light and soil requirements for the more popular ground covers.

Table 18-1

GROUND COVERS: THEIR LIGHT AND SOIL DESIRES

Shade	Sun
Carpet bugle (any soil)	Creeping phlox (any soil)
English ivy (good soil)	Juniper (light soil)
Japanese spurge (any soil)	Lantana (good soil)
Periwinkle (good soil)	Rockspray (rich soil)
Wintercreeper (rich soil)	Sedum (any soil)
	Thyme (any soil)

MULCHES   Some of the reasons for the use of mulches are similar to those mentioned for ground covers, i.e., erosion control, weed suppression, and low maintenance. It is important to realize that they are designed to do more than simply keep plants warm during the winter. Mulches are also designed to stabilize soil temperatures to prevent the freezing and thawing conditions occurring during the winter months and to help keep water in the soil by preventing surface evaporation.

To be effective, a mulch can be any loose organic matter applied to the soil after the ground has become frozen. The most common ones will be mentioned in the ensuing paragraphs.

Bark from either southern pine or western Douglas fir makes an excellent mulch, lasting for many years.

Corncobs are light in weight, easy to apply, and weed-free. In certain geographical areas they are plentiful, while in other areas they are difficult to purchase.

Shredded leaves are used by many homeowners because they are abundant, provide good protection, and also supply an ample quantity of humus to the area covered.

Peat moss is desirable because it is slow to decompose, it is clean and neat, and it is weed-free. One drawback to it is that it must be kept moist; if it is allowed to become completely dry, it may be blown away by the wind.

Pine needles, when available, form a mulch that is clean, light, weed-free, but initially may be a deterrent to certain plants because it is slightly acidic. **Over time, the acidic nature of the pine needles decreases.**

Salt hay is long-lasting, clean, and light in weight. Its desirability is due to the fact that it can be reused and that it does not mat down. It is excellent as a cover for a newly seeded lawn area.

Sawdust may be too acidic for some plants but is good for a variety of fruits and vegetables. When it is used, the area must be frequently fertilized.

Straw is selected as a mulch by those who desire one that decomposes completely in one growing season. It is somewhat more effective when shredded before being used.

Wood chips or shavings are both effective materials, but they decompose slowly.

There are presently a variety of man-made materials available to be used as mulches that are not organic. Most are much more expensive than the natural organic mulches on the market. A brief description of each is presented in the following paragraphs.

Aluminum foil, when used, conserves moisture and insulates the soil. It is ideally suited for plants grown in rows, but it must be securely anchored in position.

Glass wool is fireproof, but its expense far outweighs its benefits.

Black paper and black plastic are both excellent at conserving moisture and keeping the soil warm. When used, they either control weeds or completely eliminate their growth. Both are suitable for row crops, and the plastic can be reused the following season if handled properly.

Rubber tires, ground into small pieces, could be used as a mulch and serve an important ecological function as well.

Stones help to control soil erosion. They are permanently positioned and can be either flat, round, or chips. They may not completely eliminate weed growth, but they suppress it successfully.

Plant material can be injured during the winter months for several reasons. It may be caused by the loss of moisture from excessive evaporation, by the combined action of strong winter winds and bright sunlight, or even by the excessive fertilization of plants during the early fall. Broadleaf evergreens are the most susceptible to winter injury.

Because of this potential threat, many gardeners employ a variety of materials to protect their plants. To be effective, the material selected must allow for good air circulation, yet be able to resist both wind and snow damage, and also insure that the protected plant is never deprived of moisture.

The materials commonly used are:

1. burlap attached to upright posts
2. a slatted framework, similar to a lath house
3. straw mats or evergreen branches
4. baskets or boxes for small plants
5. twine used to tie upright evergreens to prevent the snow from causing branches to droop and eventually break
6. wide boards placed in teepee fashion over the plant
7. snow fencing, especially for hedges near the street

WINTER
PROTECTION

Chapter Nineteen

Gardens

PLANNING
THE
GARDEN To achieve the satisfaction associated with the growth of any type of garden, there is specific information that the gardener must be familiar with.

Initially, the proposed gardening site must be measured completely before any plan can be drawn up. Once the site has been measured, the design of the plot must be planned out, showing exactly where each plant should be positioned. Knowing the color of bloom, the time of bloom, the flowering height, the growing space required, and the type of foliage helps to properly position the plants.

The entire garden area must be prepared before planting can occur. Organic matter, lime, and fertilizer must be added to the plot prior to its being turned over. After the complete area has been thoroughly mixed, it must be raked clean of rocks and other debris.

If the suggested area is too large to plant completely during one season, it is better to thoroughly plant a small portion of the area than to scatter the plants over the entire area. The plant material selected should be healthy and free of pests.

During the entire growing season, it is important to perform the correct maintenance techniques when they are required, such as pruning, pinching, fertilizing, watering, staking, cultivating, weeding, and thinning. To perform all these techniques properly, it is important for the gardener to know the plants selected and their individual requirements.

Since many of the plants grown involve a harvest, it is extremely important to know the correct time to perform this satisfying feat. Being unsure of when to harvest only leads to frustration and a year's efforts possibly wasted.

Annuals, perennials, and herbs are the three most popular types of plants grown in most gardens. Table 19-1 presents the common perennials, and Table 19-2 gives a listing of the most popular annuals. Table 19-3 lists the most common herbs.

Table 19-1

PERENNIALS

Scientific name Common name	Ht. in ft.	Flowering time	Propagation	Flower color	Comments
Aconitum (monkshood)	4–5 (1.2–1.5 m)	Aug.–Sept.	division	blue, bl wt	light shade
Alyssum (basket of gold)	½ (15.2 cm)	May	division	yellow	
Anemone (Japanese anemone)	2–4 (62–122 cm)	September	division	wt, pk, red	light shade
Aquilegia (columbine)	2–4 (62–122 cm)	May	seed	various	reseed each year light shade
Aster	1–3 (30–92 cm)	June	division	various	
Campanula (bell flower)	2½ (76 cm)	June	division	wt, purple	
Chrysanthemum (hardy)	2–3 (62–92 cm)	October	cuttings division	various	
(shasta or painted daisy)		June–July	division	wt, red, pk	
Delphinium	6 (1.83 m)	June	seed	various	
Dianthus (pinks)	1 (31 cm)	June	division	red, wt, pk	
Dicentra (bleeding-heart)	2–3 (62–92 cm)	May	division	pink	
Gypsophila (baby's breath)	3 (92 cm)	June	seed	white	
Hemerocallis (daylily)	2–3 (62–92 cm)	June	division	yel, orange, brown	shade
Heuchera (coral-bells)	1–2 (30–62 cm)	June	division	wt, pk, red	
Hosta (plantain lily)	1–2 (30–62 cm)	Aug.–Sept.	division	wt, lav	shade
Iris (several types)	2–3 (62–92 cm)	June	division	various	shade
Lavandula (lavender)	1½ (46 cm)	June–frost	seed, div.	lavender	
Linum (flax)	1–1½ (30–46 cm)	June	seed	blue	some varieties are evergreen
Lupinus (lupine)	3–4 (92–122 cm)	July	seed	various	give room to spread
Lythrum	3–4 (92–122 cm)	July–Sept.	division	pk, purple	hardy
Paeonia (peony)	2–3 (62–92 cm)	May–June	division	wt, pk, red	hardy, easy to grow
Papaver (oriental poppy)	3 (92 cm)	June	division	various	

Table 19-1 (continued)
PERENNIALS

Scientific name / Common name	Ht. in ft.	Flowering time	Propagation	Flower color	Comments
Phlox	1 (3l cm)	May–June	div., cuttings	wt, pk	some shade
Platycodon (balloon flower)	2 (62 cm)	July	seed, div.	blue, wt	
Primrose	½–3 (15–92 cm)	May	seed, div.	various	some shade
Salvia	3 (92 cm)	June	seed	wt, pk, lilac	
Stokesia (Stokes aster)	1 (31 cm)	July	seed, div.	blue, wt	
Violas (violet)	½ (15 cm)	April–May	division	violet, wt purple	

Table 19-2
ANNUALS

Plant name	Height in inches	Flower color	Comments
Ageratum	6–24 (15–60 cm)	wt, blue, pk	can stand shade, can be pinched
Aster	9–36 (22–90 cm)	various	
Bachelor's button (cornflower)	36 (90 cm)	blue	easy to grow
Browallia	12–18 (30–45 cm)	wt, blue	can be pinched; hanging basket
Calendula	9–18 (22–45 cm)	yellow, orange	easy to grow; can be pinched
Calliopsis	12–30 (30–76 cm)	yel, or, br, red	can stand shade and poor soil; has a short blooming period
California poppy	6–12 (15–30 cm)	various	easy to grow; difficult to transplant
Candytuft	6–12 (15–30 cm)	wt, red, pk, lav	short blooming period
Carnation	8–18 (20–45 cm)	various	can be pinched
Celosia	12–36 (30–90 cm)	various	do not pinch
Coleus	4–12 (10–30 cm)	variegated leaf color	flowers insignificant
Cosmos	36–72 (0.9–1.8 m)	various	
Fuchsia	8–18 (20–45 cm)	pk, red, purple, wt	give ample water; can take shade; hanging basket
Impatiens	9–18 (22–45 cm)	various	do not pinch; shade lover; hanging basket
Larkspur	12–30 (30–76 cm)	wt, red, pk, blue	difficult to transplant
Lobelia	4–8 (10–20 cm)	wt, pk, blue	can stand shade
Lupine	24–36 (60–90 cm)	pur, pk, blue, yel	can stand shade; difficult to transplant
Marigold	6–18 (15–45 cm)	yellow, orange	can grow in acid soil
Nasturtium	9–15 (22–38 cm)	yellow, orange	difficult to transplant
Petunia	6–12 (15–30 cm)	various	can stand shade; can be pinched

Table 19-2 (continued)
ANNUALS

Plant name	Height in inches	Flower color	Comments
Phlox	6–12 (15–30 cm)	various	can be pinched
Poppy	18–36 (45–90 cm)	wt, red, pk	short blooming period; difficult to transplant
Portulaca	4–5 (10–13 cm)	various	can stand hot, dry conditions
Salvia	24–42 (0.6–1.05 m)	wt, red, pk	can be grown hot and dry
Snapdragon	8–36 (20–90 cm)	various (no blue)	can be pinched
Statice	12–24 (30–60 cm)	various	flowers are long-lasting
Stock	12–24 (30–60 cm)	various	do not pinch
Strawflower	24–36 (60–90 cm)	various	short blooming period; slow-growing; flowers long-lasting
Sunflower	36–96 (0.9–2.4 m)	yel, or, br	difficult to transplant
Sweet pea		various	difficult to transplant
Verbena	6–8 (15–20 cm)	pastel colors	slow-growing; can be pinched
Zinnia	9–36 (22–90 cm)	various	can be grown in hot, dry areas; can be pinched

Table 19-3
MOST COMMON HERBS

Name	Plant type	Propagation	Uses	Comments
Anise (Pimpinella anisum)	annual	seed	food, perfume, medicine	must be thinned
Balm (Melissa officinalis)	perennial	seed	medicine, food perfume	pinch top for more leaves
Basil (Ocimum minimum)	biennial	seed	seasoning, fragrance	grows in house
Borage (Borago officinalis)	annual	seed	food	
Chives (Allium schoenoprasum)	perennial	bulbs	flavoring	good edging plant
Costmary (Chrysanthemum balsamita)	perennial	root cutting	medicine, food	
Dill (Anethum graveolens)	annual	seed	for pickling, vinegars	
Horehound (Marrubium vulgare)	perennial	seed	medicine	
Lavender (Lavandula—various)	perennial	cuttings	flavoring, fragrance	needs mild climate
Marjoram (Origanum majorana)	perennial	seed	flavoring, fragrance	not winter hardy
Mint (Mentha—various)	perennial	cuttings	flavoring, fragrance	has square stem; must be controlled

Table 19-3 (continued)
MOST COMMON HERBS

Name	Plant type	Propagation	Uses	Comments
Parsley (*Petroselinum hortense*)	biennial	seed	medicine, food	
Peppermint (*Mentha piperita*)	perennial	cuttings	medicine, food, perfume	grows in house
Rosemary (*Rosmarinus officinalis*)	perennial	seed, cutting	medicine, perfume	
Sage (*Salvia officinalis*)	perennial	seed, cutting	medicine, perfume, food	
Savory (*Satureja hortensis*)	annual	seed	medicine, food	aromatic
Tansy (*Tanacetum vulgare*)	perennial	seed, division	medicine, perfume	
Tarragon (*Artemisia dracunculus*)	perennial	root cuttings	perfume, flavoring	grows in house
Thyme (*Thymus*)	perennial	seed, cuttings	medicine, perfume	good edging plant

Chapter Twenty

The Vegetable Garden

For those individuals who have never had the experience of raising their own vegetables, it can be a fulfilling and joyful one, provided that certain specific steps and guidelines are followed.

No one should start a vegetable garden without forming a plan. The area selected for the garden must receive at least 9 to 10 hours of sun per day. Insufficient light causes spindly growth and vegetables of poor quality. The garden area should not be positioned near the house or shade trees.

ORGANIZATION

Most individuals starting their first garden attempt to plant too large an area. It is much smarter to work a small area that can be easily maintained than an area too large.

Once the dimensions of the garden have been established, the length of the rows is used to help determine the amounts of seed to buy. Perusing the various seed catalogues during the winter months will help the gardener to decide upon the best variety available. The seed that is purchased must be treated against fungi and be disease resistant.

These catalogues also explain to the potential vegetable grower the most suitable planting dates for each vegetable grown in the United States; they also tell which vegetables should be started indoors.

The positioning of plants within the garden must also be understood before the plan can be considered complete. The early sown plants, like peas and spinach, must be planted near each other. Tall crops like corn, tomatoes, and potatoes should be placed at one end of the garden to avoid shading other plants.

Vine-type plants, those affected by the same insects and diseases, and plants belonging to the same family should all be planted close together in the garden. Cabbage, broccoli, cauliflower, and Brussels sprouts are all members of the same family, while tomatoes, peppers, potatoes, and eggplant belong to another plant family.

Prior to the planting of any vegetable, the garden soil should receive certain materials before it is turned over. It is accepted procedure to apply ground limestone to the surface of the vegetable garden in early spring. The actual amount necessary is determined by a soil test taken late in the fall.

If fresh manure is to be applied to the garden as a source of organic matter, it should be applied during the winter and given three to four months to decompose before it is mixed into the soil. Green manure is an excellent substitute when manure is unavailable. Winter ryegrass is sown in the fall at a rate of 1–2 lb (0.45–0.91 kg) per 1000 ft² (92 m²) and is turned under the following spring.

A complete 5-10-5 fertilizer should be spread over the complete garden area before the soil and its additives are thoroughly mixed.

If the garden area is small, the soil and its additives can be turned with a garden spade to a depth of 6–8 in. (15–20) cm). For a larger area, a rototiller is suggested. Weeds, grass clumps, and large stones should be removed at this time.

PLANTING
Sowing Indoors

As a general rule, those vegetables recommended for indoor planting should be started four to seven weeks prior to their outdoor planting date.

The quantity of a specific vegetable desired will determine which container type should be selected. For small quantities, individual peat pots, jiffy 7's, or small peat flats would be the most suitable. On a commercial scale, where large quantities are needed, flats are the most convenient to use.

The soil mixture for indoor sowing should be light and porous, and once seeding has been accomplished, it should be kept constantly moist. After germination has occurred, the container should be placed in a well-lighted area. If the container is left unattended during the day, make certain to keep it out of the direct sun.

Broccoli, Brussels sprouts, cabbage, and cauliflower can be sown indoors from March 1 to April 15. This same time interval is also recommended for cucumbers and summer squash.

Lettuce, peppers, tomatoes, and pumpkin seeds are usually sown only during the month of March, while eggplant should be sown within the first half of the month because of its very slow growth rate.

Sowing Outdoors

As mentioned in an earlier chapter, the depth to plant most seeds is approximately three times the diameter of the seed. Seeds are planted in either rows (drills) or hills in the garden. It is important for the gardener to know the correct distance between seeds in a row and also the distance between rows. This information is normally printed on every seed packet. A hill is actually a cluster of plants, and not a mound of soil. Plants sown in hills are cucumbers, squash, pumpkins, and melons. Seed should never be planted in mounds of soil, because the soil in the mound dries out much faster than when level and therefore could harm the seed.

Most seed packets have the correct thinning procedure clearly printed on the back of the packet. This garden practice is very important to the eventual yield of the vegetable, and when it is neglected, the gardener may not be successful at all with his crop. *THINNING CERTAIN PLANTS*

Thinning is done to insure an optimum yield for your gardening efforts. When it is done properly, the remaining plants experience much better growth because they receive more light, more moisture, and have additional room for expansion.

Thinning should be accomplished while the plants are small and when the soil is moist. When carrots, beets, turnips, and onions are thinned, the small removed plants are often cooked and consumed by the gardener.

In the vegetable garden, cultivation is done for weed removal, aeration of the soil, to reduce soil compaction, and to provide support for certain plants. Most gardeners use a hoe or cultivator for cultivation. *CULTIVATION*

Cultivating should be shallow to avoid damage to the roots, and is best accomplished when the soil surface is relatively dry.

When cultivation is performed every 10 to 14 days, the small weeds can be left on the soil surface and will supply organic matter to the soil.

To reduce the time spent in cultivating, some vegetable gardeners prefer to use mulches. The mulch is applied to prevent the wind from drying out the soil, to reduce or eliminate weed growth, and to shade the soil and prevent wide fluctuations in soil temperature. When black plastic is used in the garden, it is positioned on the surface either just before or directly after planting. Tomato plants should be mulched once they begin to flower and display small pea-sized fruits.

When the vegetables do not receive sufficient rainfall during the week, it is necessary to water the garden. Water can be applied at any time of day, and it should be added slowly so that it percolates to a depth of at least 3 in. (3.75 cm). Care should be exercised not to overwater the vegetable garden.

As mentioned earlier, every vegetable gardener must take every precaution to minimize the harm caused by insects and diseases. This is done by making certain that both the crops within the garden and the garden area itself are constantly rotated, that resistant varieties are planted, that all seeds are pretreated, and that the garden area has been properly fertilized, to ensure the desired growth rate of each vegetable. *DIAGNOSING VEGETABLE PROBLEMS*

Table 20-1 gives the common troubles associated with vegetables, their cause, and how to control each one.

Each vegetable should be harvested when it has reached its most edible stage. For many vegetables, this statement presents no problem whatsoever, while with others, certain guidelines should be used. Most *HARVESTING*

Table 20-1

VEGETABLE PROBLEMS AND THEIR CONTROL

Identifying problem	Cause	Control
Seedlings die before maturity	damping off fungi	Sterilize soil
White fuzzy mold on leaves	mildew	Keep foliage dry; spray fungicide
Dark dead spots on leaves	leaf spot disease	Same as for mildew
Plants cut off at base	cutworm	Use paper collars
Holes in leaves	leaf-chewing insects or slugs	Spray malathion Apply metaldehyde
Leaves mottled with light and dark areas	mosaic virus	Plant resistant varieties and remove weeds near plants

legumes, for example, should be removed from the parent plant before the seed has become completely mature or has filled the interior of the pod completely. Many of the fleshy vegetables are sweeter if picked during the middle of the day when the plant is actively engaged in photosynthesis.

SPECIFIC VEGETABLES AND THEIR REQUIREMENTS

The following tips should be extremely helpful to anyone growing vegetables in a temperate climate. For warmer areas, the performance of a specific procedure (i.e., sowing) should commence three to six weeks earlier than the dates suggested.

Asparagus

1. Sow the seed as early as possible—soak the seeds at 85°F (29°C) for two days before sowing.
2. Select a well-drained site.
3. Cover the area with hay or straw mulch.
4. Remove the mulch only during the spring.
5. Add manure or compost each fall.
6. Transplant the roots the following spring.
7. Be conservative the second year to ensure a heavy crop the third year.
8. Cut the tender spears level with the ground when they are 4 to 6 in. high until July 1.
9. Allow growth to become bushy during the summer, and trim it back to the ground in late September.

Beans

1. Sow every 10 days from May 20 on (1 in. deep).
2. Soil temperature must be 60°F (16°C) for germination.
3. The seeds will not germinate if they are planted too deep or too early in cold wet soil.

4. Do not cultivate when the plants are wet, i.e., after a rain or in the early morning. This may cause disease.
5. Plant marigolds or garlic between the rows.
6. Apply fertilizer after sowing in bands parallel to the rows (2 in. from the row).

Beets

1. Sow every two weeks from early April to July 15.
2. Beets grow best in rich, sandy loam that is free of stones.
3. Harvest when the roots are at least 1½ in. in diameter.
4. Beets can be stored for the winter in moist sand at 32°F (0° C) and 95 percent humidity for six months.
5. Sow ½ in. deep and thin to one plant every 3 in. when the plants are 2 in. high.
6. Beets can be sown indoors March 1 to be transplanted outdoors around April 15.
7. Seed germinates in 4 to 10 days [45 to 70°F (7 to 21°C) soil temperature].

Brussels Sprouts

1. Sow outdoors in seed beds from May 10 to 25 (¼ in. deep and 3 to 4 plants per inch).
2. Transplant when the plants are 5 in. tall at 2-ft intervals.
3. The plants require much moisture and nitrogen.
4. Brussels sprouts must be kept weed free and well mulched.
5. Do not cultivate too deeply.
6. Bees help by eating cabbage worms.
7. Brussels sprouts can only be planted in an area once every four years.
8. About September 15, pinch out the center stem and remove the lower leaves and stems.
9. Harvest the sprouts from the bottom of the plant upwards when they reach 1 in. in diameter.

Broccoli

1. Broccoli requires loose rich soil that retains moisture.
2. Sow seeds indoors from March to early April.
3. Transplant outdoors in late April or early May at 1½-ft intervals.
4. Sow outdoors from early May to the first week in June for a late crop—thin when plants are three weeks old.
5. Spray with Sevin dust weekly to all parts of the leaf.
6. Plant chives or garlic between the rows, also nasturtiums for insect control.
7. Plant in the same area once every four years.
8. Harvest just before the tiny buds of the head start to open.

Cabbage	1. Sow seed indoors in Jiffy 7's from February 1 to March 15.
	2. Seeds will germinate in four to five days if the soil temperature is near 80°F (27°C).
	3. Transplant outdoors in 5 to 6 weeks at intervals of 15 in.
	4. Do not cultivate too deeply.
	5. Keep well fed and watered.
	6. Rotate the growing area (once every four years in same location).
	7. Follow the pest controls mentioned for broccoli and brussels sprouts.
Carrots	1. Carrots must be grown in loose, sandy soil tilled to a depth of 12 to 14 in.
	2. Do not use fresh manure.
	3. Sow outdoors from June 1 to July 15, ¼ to ½ in. deep and 1/8 in. apart; thin to 1½ to 2 in. apart.
	4. Avoid compacting the soil in and between the rows (stand on a plank).
	5. Seeds germinate in 6 to 14 days at a soil temperature of 60°F (16°C).
Cauliflower	1. Cauliflower likes a fertile, organically rich, well-limed loam.
	2. Sow indoors from February 1 to March 15.
	3. Transplant outdoors after April 15, at intervals of 18 in.
	4. The plants need slow, steady growth; they are greatly affected by weather fluctuations (it is best to sow and transplant at weekly intervals).
	5. Tie heads before they reach the size of a teacup.
Celery	1. Start seeds indoors in Jiffy mix about February 25.
	2. Transplant 10 to 12 weeks after indoor sowing, but not before June 1.
	3. Keep the plants well watered and fertilize them every two to three weeks.
	4. Mulch the plants with clean straw, adding more as the plants grow.
	5. Hand pick the celery worms and control aphids with Rotenone dust.
Corn	1. Add 5-10-5 fertilizer to the soil area before sowing.
	2. Sow at least four rows of each variety, ½ in. deep, and keep the rows 2½ ft apart for wind pollination.
	3. Thin to two sprouts per hill.
	4. Once the stalks reach 18 in. in height, apply Sevin every five days into the axils of each leaf (this controls corn borers); increase the frequency to every two to three days when the silk appears.
	5. Corn borers can be avoided by planting after June 15.
	6. Remove and destroy any large black puffy balls that appear during hot, dry weather (corn smut prevention).

1. Sow indoors from April 15 to May 1 in peat pots (three to four per   *Cucumber*
   pot).
2. Transplant outdoors after June 10.
3. Sow directly outdoors two weeks after the last frost—i.e., June 1
   planting, four to six per hill.
4. Thin outdoor-sown hills to three to four plants when they are 6 to
   8 in. tall.
5. When transplanting, keep the indoor-sown peat pots well watered,
   dust with an insecticide, and do not tear out the bottom of the pot.
6. Mulch the planted area with black plastic.
7. Radishes, nasturtiums, or marigolds help keep beetles away.

1. Sow indoors from March 1 through April 1, and keep very warm—   *Eggplant*
   the soil temperature should average 85°F (29°C).
2. Reduce the moisture prior to transplanting.
3. Transplant outdoors once the danger of cool nights has passed.

1. Head lettuce will grow in average soil but does much better in well-   *Lettuce (Head)*
   fertilized, relatively moist soils.
2. Most insects can be washed off.
3. Sow indoors from February 15 to March 15, and transplant outdoors
   in April after first hardening off the plants.
4. Sow outdoors April 1, ¼ in. deep, two seeds per inch; thin to leave
   one head per foot.
5. Heads mature during July and August.

1. Shell the nonroasted nuts and sow from May 15 to June 1 in warm   *Peanuts*
   sandy soil having a southern exposure.
2. Sow nuts 3 in. apart and 1½ in. deep.
3. Keep plants well cultivated.
4. Flowers form, become pollinated, and root themselves in the soft
   ground, giving rise to the clusters of peanuts beneath the soil.
5. Dig up the nuts before the first frost and hang them up to dry and
   cure.
6. Roast at 300°F for one hour with frequent stirring.

1. Sow 2 in. deep from February 15 to May 15 (10 seeds per foot).   *Peas*
2. Seeds germinate in five to eight days when the soil temperature is
   near 60°F (16°C) and there is ample moisture.
3. Peas prefer rich, low nitrogen, sandy soil.
4. Provide some form of support for the vines.
5. Harvest daily when the pods become well filled.
6. Choose disease-resistant varieties.
7. Plant in the same area only once every three years.

Peppers   1. Sow indoors from February to March 15 in Jiffy mix at a soil temperature of 70°F (21°C). Plant ¼ in. deep.

2. Transplant outdoors two months after germination occurs (cutworm damage can be prevented by waiting until June 1; plant at 18-in. intervals.

3. When plants are 5 in. tall, pinch the outer terminal tip to force branching.

4. When fruits appear, plants should be staked.

5. Weed control is achieved by either mulching or shallow cultivation.

6. Cold June nights cause blossoms to drop.

7. Fertilize before transplanting and again when plants are 12 in. tall.

Potatoes   1. Plant pieces of resistant seed potatoes containing two to three eyes at a depth of 3 to 4 in. in sandy, acid soil.

2. Fertilize the soil before planting and again one month later.

3. Dig up the potatoes before the first frost.

Pumpkins   1. Sow three to four seeds per hill about June 1.

2. Remove all but the strongest plant in the hill when they are 2 to 3 in. tall.

3. Mulch heavily and provide ample moisture.

4. The addition of well-rotted manure or compost ensures large-sized fruit.

5. Remove all but one blossom from each vine if very large pumpkins are desired.

6. Harvest before the first frost, making certain to have at least 3 in. of stem attached—once accomplished, place the pumpkins in the sun to harden.

Radishes   1. Sow seed ½ in. deep as soon as the ground is workable.

2. Seed germinates in five days in well-drained soil.

3. Thin 2-in. tall seedlings, leaving one plant per inch.

4. Keep plants moist, but do not overwater.

5. Do not sow in old cabbage-growing areas since maggots from the cabbage will appear.

6. Fertilize the soil before planting.

7. Harvest when radish tops appear above the ground.

Spinach   1. Sow seed as early as possible until early May, 1 in. deep and 3 in. apart in the row.

2. Thin 3-in. tall seedlings to one every 12 in.

3. Fertilize before planting and when thinning is done.

4. A well-drained, richly organic, and well-limed soil is desirable.

136

5. Harvest when the leaves are large enough to use by cutting the plant at its base.

1. Sow indoors in Jiffy 7's about April 15 or directly outdoors about June 1.   *Squash (Summer)*
2. Sow four to five seeds per foot or six per hill outdoors, ½ in. deep, and thin to the three strongest plants when they are 3 in. tall.
3. Fertilize when thinning and when vines begin to spread.
4. Plants will continue to produce if fruits are constantly picked.
5. Harvest as the blossom ends drop off the fruit until they reach 8 in. in length.

1. Sow outdoors June 1.   *Squash (Winter)*
2. Harvest when the leaves start to turn brown and before the first frost.

1. Sow indoors from February 15 to April 1, attempting to maintain a   *Tomatoes*
soil temperature of 80°F (27°C).
2. Transplant outside after the danger of frost is over.
3. Set plants deep in the soil (to the bottom leaves) and 2 to 3 ft apart.
4. Hybrids need more fertilizer, need to be staked, and need to have their suckers removed when small.
5. Fertilize when planting outdoors, during flowering, and every week after the fruits appear.
6. Mulching is beneficial and is best done when the flowers appear.
7. Harvest when the entire fruit is red.

Chapter Twenty-One

Home-grown Fruits

The appeal of home-grown fruit is normally greater than the fruit purchased in the market, because it usually has a better taste and is generally more ripe than the market fruit. This desirability continues only as long as the home gardener follows a vigorous spray program for insect and disease control, and properly prunes his plants when it is required.

For most home sites, the planting of standard-size fruit trees is impractical, because each tree must be positioned some 30 ft (9.2 m) from the next tree. Dwarf fruit trees are more practical, since they require a spacing of only 5 ft (1.53 m) between trees. It must be understood that the purchase of a healthy tree is no guarantee that it will yield good fruit. It is necessary for the grower of any fruit tree to know exactly what must be done and when to do it.

Dwarf fruit trees, cane fruits, strawberries, and grapes are practical for even the very small garden. They require little room and can be cared for easily and inexpensively.

All fruits should be planted on fertile, well-drained soil, should receive abundant sunlight, should be periodically sprayed to protect against specific insects and diseases, and should be pruned annually during the dormant season.

The balance of this chapter concerns the important considerations and facts associated with the most popular fruits grown.

TREE FRUITS
Apple

After two years of growth, most varieties of apple are ready to bear fruit. When fertilized properly, the bearing trees should yield 8–10 in. (20–25 cm) of terminal growth in one year. Two-thirds of this growth is removed while the tree is dormant (February and March) leaving four to five buds per stem. The terminal bud will produce a new shoot, while the other buds will carry the fruit. Apples must be provided with good air circulation and should never be planted in low spots. Mulch should be placed beneath each tree.

Most varieties should be pollinated with pollen from another compatible variety. Insect pests include scale, aphids, codling moth, and leaf rollers. Scab, cedar rust, and fire blight are the three most common diseases affecting apple trees. The spray program for all tree fruits will be presented in table form at the end of this section.

Sweet or sour cherry trees should bear fruit in three to four years after planting. Sweet cherry trees will not produce cherries unless they are cross-pollinated with another variety, whereas sour cherries do pollinate themselves. Beneath each tree the soil must be cultivated three to four times during the season and before the harvest occurs. After fruiting occurs, a cover crop should be sown. Birds may be more detrimental to the cherry crop than the pests. The common insects that attack cherries are scale, plum curculio, and fruit fly, and the diseases are botrytis blossom blight, brown rot, leaf spot, and powdery mildew.

Peach trees are the least hardy of all the fruit trees grown in the temperate zone. Trees should bear fruit after two to three years. Ideal growth per year is 12 in. (30 cm) and is usually attained by applying ½ lb (2.27 g) of nitrogen fertilizer per tree. Since peach trees are somewhat fragile, they should be pruned lightly for the first four years of their growth, and more severely as the tree increases in age. In June, the fruit should be thinned to leave one fruit for every 50 leaves; if this is not done, whole branches laden with fruit would collapse.

Sanitation measures are extremely important for the suppression of pests, and it is important to spray the trunk and all the branches each time the fruit and foliage are sprayed. Ryegrass should be sown under the trees and should be disced rather than turned under.

The insects affecting peach trees are the tarnished plant bug, peach scale, plum curculio, oriental fruit moth, peach tree borer, and European fruit lecanium scale. Diseases are leaf curl, mildew, scab, and brown rot blossom blight.

Pear trees require light but regular pruning to prevent the tree from becoming too dense. Any sod layer beneath the trees must be eliminated by cultivation, mulching, or the use of herbicides. Pears are less hardy than apples, and cross-pollination with another variety is suggested.

The diseases affecting pear trees are fire blight, scab, fabraea spot, and sooty blotch. The insects attacking the trees are varied and numerous and include the leaf blister mite, psylla, European red mite, pear plant bug, tarnished plant bug, aphids, leaf hoppers, and plum curculio.

It is important to plant more than one variety of plum to ensure cross-pollination. Because they are early to bloom, they are easily injured by frost. Mulching or discing the sod cover under each tree is very satisfactory to the overall growth of each plum tree.

The red-banded leaf roller, plum curculio, eye spotted bud moth, European red mite, and lecanium scale are the insects that endanger the plum crop, while the diseases causing damage are leaf spot, brown rot blossom blight, and black knot.

FRUIT TREE
SPRAY
PROGRAM
Many homeowners are frightened away from raising fruit trees because of their lack of knowledge of just how involved the complete spraying program is. They envision that every fruit tree must be sprayed differently than all the others, that a complex battery of chemicals is required to perform the spraying, and that great expenditures of time are needed to complete the various operations. These are incorrect assumptions.

Once they are informed that there are only three different types of solutions required for the spraying of all fruit trees and that the frequency of application for the homeowner is approximately once every two weeks, weather permitting, their mouths begin to water for the taste of home-grown fruit.

The first type of spray applied is the "dormant spray." Commercially, dormant sprays are available as either a lime-sulfur mixture, an oil spray, or a dinitro compound. Any of the three can be applied to dormant fruit trees from March through April before the buds start to swell, provided that the daytime temperature is relatively warm. The design of the dormant spray is to kill those insects that overwinter on the fruit trees and prevent them from hatching and causing serious harm to the trees.

The second type of spray applied is a fungicide applied only when the blossoms are open. The one most commonly used is Captan (50% wettable powder). No insecticide must be used at this time, and any other fungicide selected must not harm the pollen-carrying bees.

The third type of spray applied to fruit trees will be referred to as a multipurpose spray mixture or "M–P." This mixture contains the following ingredients in one gallon (3.75 l) of water:

Malathion (25% wettable powder)	2 tablespoons
Methoxychlor (50% wettable powder)	3 tablespoons
Captan (50% wettable powder)	3 tablespoons

This mixture is applied to all fruit trees at specific stages of their bud, flower, and fruit development, i.e., preblossom, petal fall, and cover sprays. (See Table 21-1.)

Preblossom occurs when the buds begin to show color, but before the petals have begun to unfold. Petal fall is the interval of time immediately

Table 21-1

A TYPICAL SPRAY PROGRAM FOR FRUIT TREES

Stage of development	Material to be used
Dormant or delayed dormant	Lime-sulfur, oil, or dinitro
Preblossom	M–P spray
Blossom	Captan or other suitable fungicide
Petal fall	M–P spray
First cover	M–P spray
Consecutive cover sprays	M–P spray

after the last petal has fallen from the flowers. Cover sprays are applied at successive intervals of 10 to 14 days after the petal fall has been applied and continued to within a month of the expected harvest. Of the three, the most important are the preblossom and petal fall sprays.

Canes of this fruit are biennial, requiring two years of growth before they bear fruit. New plants are topped at a height of 3 ft (90 cm) when planted and are cut back to 1½–2 ft (45–60 cm) the second spring; once the canes bear the fruit, they are removed. Blackberries grow best in a soil well fortified with organic matter and require excellent drainage. Cultivation should be shallow, and suckers must be removed as soon as they appear. They are less hardy than raspberries and should not be allowed to overrun their planting area but should be kept in rows no more than one foot wide. *VINE AND CANE FRUITS Blackberry*

These plants prefer to be grown in full sunlight, with adequate ventilation, and in a sandy, acid soil. They are self-fertile, but the fruit yield may be greater if cross-pollination is provided by another variety. After three years of growth, all weak shoots and suckers must be removed. Pruning is done after the leaves have fallen; if too much foliage is left, small, inferior fruit will result. Blueberry plants are subjected to attack by both insects and diseases, which may necessitate their being sprayed. Sanitation is usually the method employed to rid the plants of the undesirable pests. This is accomplished by removing the affected parts and burning them. These plants prefer mulching to cultivation; any sawdust available can be used as the mulching material. *Blueberry*

A variety of soils can be used to grow grapes, but they all must have a high organic content. When planted, each cane must be pruned, only two buds per cane being left. The following year, 8 to 12 buds are left on each of the previous year's canes. A good mature grapevine can support approximately 50 buds on four to five good strong canes. Sanitation is extremely important in the culture of all grape varieties, and mulching the plants every second year of growth will help maintain the desired organic content of the soil. It is advisable to cover the fruit to repel birds. Because of the large size of most grape leaves, all grapes are extremely sensitive to 2, 4–D. When the grape buds begin to swell, each vine should receive ⅓ lb (151 g) of nitrate of soda and 2 lb (0.91 kg) of potassium sulfate. After the second year of growth, all grape vines will need some form of support. *Grape*

Like the blackberry, these plants are also biennial. New plants are obtained each year from either suckers or tip layering. They are cut back to 12 in. (30 cm) in height when planted and cut back the second year to a height of 3½–4½ ft (1.06–1.38 m). If both red and black varieties are grown, they must be separated because of the mosaic virus, which is spread by aphids from the wild to the tame raspberries, resulting in dried-up fruit and stunting of the plants. These plants should be frequently cultivated but never deeper than about 3 in. (7.5 cm) until the blossoms form. *Red Raspberry*

Strawberry   Blossoms should be removed from the first-year plants to insure a good second-year crop. It is also important to remove all but six runners from each plant. Strawberries need protection from the birds and winter conditions. A mulch of straw or salt hay is applied before the ground freezes. They should not be planted where tomatoes, potatoes, or peppers were formerly planted, because they will die from the remaining verticillium wilt. During their bearing year, strawberry plants must be frequently watered.

Chapter Twenty–Two

Nursery Management

A plant nursery can be described as any commercial enterprise in- *NURSERY* volved in the production and distribution of any or all of the following plant *GROWN* types: *PLANTS*

Ornamental trees and shrubs

Fruit plants, both vine and tree types

Perennials, both flower and vegetable (asparagus, rhubarb)

Roses

Christmas trees

Annuals and biennials

The necessary operations that must be performed, in any nursery, no matter what plant types are grown, fall into two separate categories, i.e., (1) those involved with the purchase and resale of plants, and (2) those involved with the maintenance of existing plants.

In either the purchase or resale of plant material, the operations requiring physical labor are

1. *planting* the material in the ground or in containers,
2. *transplanting* a plant from a container to a nursery bed or the reverse,
3. *root pruning* to initially prepare the plant for its eventual removal (digging) from the nursery row, and
4. the *packaging* of plants for sale.

In addition to the necessary manual labor tasks, nurseries must (1) keep up-to-date *inventories* of existing stock, (2) keep *cost accounting records* to determine profits and losses, and (3) *grade* their stock to ensure that they are receiving a fair and just price for their plant material. Without these records, it is entirely possible that many nurseries would be selling their grown plants at prices lower than they should command.

The maintenance of all nursery-grown plant material encompasses the following operations:

1. *Watering.* This must be performed daily for those plants grown in containers or positioned out of the ground and wrapped in burlap. Plants growing in the nursery are usually watered only during extremely dry spells.

2. *Fertilization.* The frequency of application is related to the type of plant grown and whether or not it is planted in the nursery or offered for sale in a container.

3. *Pest control.* With the knowledge that their entire business can be completely closed down to the public whenever a state or federal inspector discovers a serious problem on a given plant, all nurseries are forced to maintain a weekly, routine program of pest control.

4. *Cultivation.* This function is necessary to ensure an optimum growth rate for all plant types and is also done to limit weed growth and to aer .te the adjoining soil.

5. *Pruning.* To prepare each plant for its eventual sale, the nursery employees must know when and how to trim and shape their plants.

6. *Winter protection.* This is necessary for those nursery-grown plants that are not completely hardy in their present location. Years of experience have usually been the best form of educating nurserymen to which plants require protection.

7. *Soil enrichment.* Since many nursery-grown plants remain in a specific location for several years and are constantly depleting the soil during this time span, cover crops are frequently planted between the plant rows to help maintain a certain level of organic matter and also to suppress weed growth.

Nurseries market their trees and shrubs in one of three ways:

1. *Balled and burlapped.* This method is used for all large trees and shrubs with an upright habit. It consists of wrapping burlap around the plant's root ball. With the use of large machinery, some nurseries prefer to transplant their extra-large trees during the winter months while the trees are in a dormant state.

2. *Container-grown.* An advantage to the use of containers is that it provides the nursery with year-round sales possibilities. The two most popular types of containers used are (1) large metal cans with cutouts at the bottom for drainage, which require removal of the plants from the cans before planting, and (2) large peat moss composition containers designed to decompose within months after they and their contents have been planted. Some large nurseries blend the ingredients for their container-grown plants in cement mixers adapted for this purpose. The ingredients usually consist of soil, cinders for drainage, and peat moss for water retention.

3. *Bare root.* This method is most commonly used by mail order nurseries to supply their customers with plant material at prices usually
slightly below retail. The roots are completely surrounded with
either moist sphagnum moss or peat moss and tightly packed in an
airtight, waterproof package. This is necessary to ensure that the customer receives a healthy, viable plant no matter how long the delivery takes.

In addition to the sale of plant material and plant-related supplies,
certain nurseries provide their customers with additional services. These
include:

1. *Total landscape service.* They provide their customers with a suitable
   landscape plan drawn by one of their architects and with landscape
   crews to do the actual planting.
2. *Home maintenance service.* This service would include lawn mowing,
   tree and shrub pruning, tree repair, and the planting and upkeep of
   flower gardens.
3. *Problem-solving service.* Some nurseries have highly trained, knowl-
   edgeable employees whose sole purpose is to answer all the gardening questions posed by their customers.

The total nursery business is divided into four different types of en-
terprises. They are:

1. *Wholesale.* They are basically propagators, who grow a limited
   number of plant types, which they supply to the retail trade. Most of
   their plants are small in size, and are referred to as "lining-out
   stock," since they have not attained a saleable size and must be
   planted in nursery rows until they become the desired size.
2. *Retail.* They purchase small plants from various wholesalers and grow
   them in nursery rows until they become large enough to resell them
   to the public. Many retail operations provide their customers with
   both a landscape design service and a planting service.
3. *Mail order.* Most of these suppliers of nursery materials usually grow
   very few plant types themselves and are dependent upon other
   wholesale growers and suppliers for the bulk of the items offered in
   their catalogues. Because their plants are sent through the mail, they
   are usually small in size and are packaged "bare root."
4. *Garden center.* The intent of most garden centers is to provide their
   customers with a complete range of gardening needs. Some combine
   the sale of nursery plants with a floral design shop in an attempt to
   both aid and maintain their customers. Other garden centers are basically supermarket operations that provide small-size plants at prices
   usually below retail. This type of enterprise must attempt to completely sell out their entire stock, because they have no nursery area
   to overwinter the plants not sold.

Chapter Twenty-Three

Greenhouse Management

The primary concern of the average consumer when purchasing either cut flowers or a potted plant is usually one of cost. His floral preference may be drastically altered but not his prearranged price. He takes the presence of the plants for granted and is totally unaware of all those factors that had ,o be properly controlled to insure production.

Proper greenhouse management is a day-to-day challenge confronting the grower of any plant grown indoors. He must thoroughly know and understand the needs of each species of plant he grows and know how to manipulate its environment to satisfy these needs.

PLANT GROWTH CONSIDERATIONS

Listed below are all the factors that the grower must consider when growing any plant.

Propagation. The grower must know the most effective method of obtaining the plants to be grown. Some plants can only be started by sowing seeds, whereas others are readily obtained from stem or leaf cuttings. It may be more economical for the grower to purchase rooted cuttings from a reliable commercial propagator than to root his own.

Soil. One advantage, common to all growers of greenhouse plants, is that the soil can be altered to fit the plants grown. It is possible to increase the soil's organic content without seriously hampering its drainage ability. All growers realize the importance of pasteurizing their soil before it can be used to support the growth of any plant material.

Soil pH is another consideration for growers. For optimum growth for each plant to be achieved, the pH preference of each plant must be known, and the soil must be altered by the grower to meet the pH demands of the plant.

Temperature. The majority of plants grown in the greenhouse can tolerate wide fluctuations in daytime temperatures, but most prefer temperatures between 60 and 80°F (15 and 26°C). Higher daytime temperatures cause plants to lose their moisture through transpiration faster than the roots can replenish it by absorption. Prolonged exposure to excessive heat will cause dehydration, a condition from which most plants are incapable of recovering. Shading and misting the foliage both help to avoid plant dehydration.

Each plant grown in the greenhouse has its own specific nighttime temperature. In most cases, the growers divide the plants grown into three categories: 50°F (10°C), 60°F (15°C), and 70°F (21°C). For economic reasons, the night temperature in a specific greenhouse may be kept 3–4°F (1.6–2.2°C) below the desired temperature, e.g., 46°F (8°C) instead of 50°F (10°C) without any marked ill effects. However, poinsettias grown experimentally at 10°F (5.5°C) intervals from 50°F (10°C) nighttime temperatures displayed wide variations in bract color. Those grown at 50°F (10°C) were an intense red, whereas those grown at 80°F (26°C) were a bleached-out red color, and the plants were unsaleable.

Watering.   The various methods of watering plants in the greenhouse have already been discussed in the chapter on watering. Each grower must determine for himself which method is the most efficient and economical to use.

On bright days while the temperature is rising, plants with smooth foliage often are sprayed with a high-pressure hose. This procedure is known as *syringing* and is done to increase the humidity within the greenhouse and around the plants, resulting in a decrease in the plant's evaporation rate, as well as to provide an alternate control to a red spider mite infestation.

Fertilization.   Most growers adopt a routine schedule of fertilizing their plants. Most pot-plant growers use a highly concentrated, slow-release granular fertilizer that may suffice for the entire growing period. Cut-flower growers usually make liquid applications of fertilizer on a regular basis. The method of watering is frequently used as the determining factor in the frequency of fertilization, since some watering systems cause far more leaching of fertilizer than do others. As cut flowers near the flowering stage, the frequency of fertilization is usually increased.

Insect and disease control.   Growers must be able to identify the characteristic signs of disease and insect infestations and must know which chemicals must be applied to the plants to achieve control. To keep problems associated with pests to a minimum, every grower should grow only healthy plants, should keep all growing areas clean and sanitary, and should maintain a routine spray schedule with pesticides to keep possible infestations from becoming serious.

Humidity.   The humidity level in the greenhouse must be kept high enough that the water lost by the leaves due to transpiration is *less* than the moisture intake by the roots of the plants. Most plants grow well in humidity levels between 30 and 50 percent. High levels of humidity can be serious during the winter months, because when high humidity is coupled with dark, overcast days, the possibility of a bacterial or fungal infection is greater. In contrast, on bright winter days with the heat on, humidity levels in the greenhouse may be too low. On bright, warm, summer days, it will be necessary for the grower to raise the humidity level by syringing either the plants or the floors and walls of the greenhouse.

All growers are aware of the fact that excessive moisture can be decreased by either ventilation or by increasing the temperature.

Ventilation. Proper ventilation provides relief from the sun's intense rays, helps to maintain the desired growing temperature, allows for the removal of air pollutants, and is beneficial in the prevention of disease organisms. Growers all realize that the ventilators should be opened in the morning as the temperature rises and closed in early afternoon to conserve heat. On cold days, the vents should be opened only a crack, and during periods of high winds, only those ventilators opposite the direction of the wind should be open.

Spacing. For commercial growers, this factor is very important, because it enables them to receive maximum production from the available growing space. This proper distance must be determined in advance of planting and is based on the type of plant grown. For example, single-stem chrysanthemums are planted at a different spacing interval than pompon-type chrysanthemums.

The types of plants grown in greenhouses can be divided into five separate categories. Some commercial establishments grow only one type, whereas others may raise all five. They are (1) cut flowers, (2) potted plants, (3) foliage and exotic plants, (4) bedding plants (annuals and perennials grown from seed and sold only in the spring), and (5) bulbous crops.

Besides the customary glass greenhouse, plants can be grown in the following structures:

1. *Plastic greenhouses.* Most of these are not as permanent as glass, but they are much less expensive. They may be covered with either polyethylene plastic or fiberglass.
2. *Cold frames.* These are outdoor plant beds that are covered with glass window frames and are used for spring seedlings and for certain methods or propagation.
3. *Hot beds.* These are cold frames with a source of heat.
4. *Cloth houses.* These structures are enclosed with cloth to keep the insect population down and are used during the summer months as a means of increasing production.

CUT FLOWERS GROWN IN THE GREENHOUSE

The cut flowers most commonly grown commercially in the greenhouse will be presented in outline form in the following pages. For simplicity, they will be presented in alphabetical order according to their generic name. The topics covered for each will follow a definite pattern so that a quick referral can be made at a later time for specific information.

Flowering season: winter *Anemone*
Time to mature: 3 to 5 months; seeds take one year
Growing temperature: 40°F (4°C)
Propagation: division of plants in the fall, or from seeds sown during the summer
Special treatment: Water should be withheld after flowering, and the plants should be stored dry and cool through the summer months to be repotted in the fall.

Flowering season: winter through spring *Antirrhinum*
Time to mature: 5 months *(Snapdragon)*
Growing temperature: 50–60°F (10–15°C)
Propagation: seeds sown during the summer months
Special treatment: Long days hasten flowering but reduce the quality, whereas short days give the reverse response. They can be grown the year round by proper selection of varieties.

Flowering season: all year with photoperiod adjustments *Chrysanthemum*
Time to mature: This is dependent upon the variety grown. (Plants *(C. morifolium)*
are classified from 6- to 15-week varieties from the time short days start until flowering occurs).
Growing temperature: 60°F (15°C)
Propagation: stem cuttings at any time during the year
Special treatment: The size of the plant desired determines the length of the vegetative period or long days. During natural short days, plants must be lighted during the middle of the night to keep them vegetative and prevent flower buds from being initiated. Pompon varieties are usually pinched to induce branching.

Flowering season: continuous but less during the winter months *Dianthus*
Time to mature: 6 months *caryophyllus*
Growing temperature: 50°F (10°C) *(Carnation)*
Propagation: stem cuttings from winter through early spring
Special treatment: Carnations prefer high light intensity for maximum flower production. Newly planted cuttings should receive at least two pinchings; one of the terminal tip, and the second to the resulting laterals. All axillary flower buds must be removed from the flowering stem, only the terminal bud being left. Flowering should occur in 10 to 20 weeks after the two pinchings have been made.

Flowering season: late winter *Freesia*
Time to mature: 3 to 4 months for corms, and one year for seeds
Growing temperature: 45–55°F (7–13°C)
Propagation: Corms are planted in the fall, whereas seeds are sown during the spring.
Special treatment: The mature corms should be dried off after flowering and kept cool until the fall, when they are replanted. Short days increase the quality of the flowers.

Gladiolus

Flowering season: spring
Time to mature: 2 to 3 months
Growing temperature: 65–70°F (18–21°C)
Propagation: corms planted during the winter
Special treatment: The corms must be cured for two weeks at 80°F (27°C) after flowering has been completed; then they are stored at 45°F (7°C) until the following planting season.

Iris (Dutch type)

Flowering season: winter through spring
Time to mature: Eight to 10 weeks of forcing time is needed after the bulbs have been stored for at least 6 weeks at 45°F (7°C).
Growing temperature: 50°F (10°C)
Propagation: bulbs planted in the fall and stored at 45°F (7°C) until the foliage is 2 in. tall
Special treatment: High temperature and poor root systems cause the bulbs to blast (not form flowers). Precooled bulbs need not be stored at 45°F (7°C).

Lathyrus odoratus (Sweet pea)

Flowering season: winter through spring
Time to mature: 2 to 4 months
Growing temperature: 50–55°F (10–13°C)
Propagation: Seeds are sown during late summer or early fall.
Special treatment: Discard the plants once they have produced their fragrant flowers.

Mathiola incana (Stock)

Flowering season: winter through spring
Time to mature: 4 to 6 months
Growing temperature: 45–60°F (7–15°C)
Propagation: Seeds are sown during the summer or fall.
Special treatment: The plants should be discarded after the fragrant flowers have been produced. Long days hasten bud development at 60°F (15°C), but growth becomes weak.

Narcissus (Daffodil)

Flowering season: winter through spring
Time to mature: 3 to 6 weeks after cool storage
Growing temperature: 55–60°F (13–15°C)
Propagation: Bulbs are planted in the fall and stored for a minimum of 10 weeks at 46–48°F (8–9°C).
Special treatment: The flowered bulbs may be planted in the garden, but can not be forced again. In the garden they normally take two years to recover from the forcing before they flower again.

Tulipa (Tulip)

Flowering season: winter through spring
Time to mature: 10 to 15 weeks after planting
Growing temperature: 60°F (15°C)
Propagation: Bulbs are planted in the fall.
Special treatment: The bulbs must receive 8 to 14 weeks of storage at

46–48°F (8–9°C) before they can be forced into bloom. Bulbs are either discarded after flowering or are placed in the garden.

Potted plant growers must be totally familiar with all the necessary procedures needed to properly schedule their plants for a specific holiday. Using common names for the plants grown, the following pages will give an alphabetical presentation of the information required to have plants ready for Christmas and Easter or Mother's day.

Azalea. Bud initiation occurs during the summer months. Bud development requires 4 to 6 weeks at a temperature of 45–48°F (7–9°C) until November 5. The remaining 4 to 6 weeks are used to force the plant into flower at a night temperature of 60°F (15°C).

Christmas begonia. Leaf petiole or stem cuttings are taken during the winter and early spring to produce a saleable plant in the month of December. These plants need short-day conditions to flower for Christmas.

Christmas cactus. This plant requires short days from September 1 on until it flowers and should be kept on the dry side during the months of September and October.

Chrysanthemum. The rooted cuttings should be placed in a pot of the desired size. They may be grown as single-stem plants, or they may have the terminal tip pinched so that each stem will yield several flowers. The cuttings are lighted for 1 to 4 weeks from 10 P.M. to 2 A.M. for their vegetative period. This is followed by the specific number of weeks of short days required by the variety grown, 8 to 12 weeks. For an 8-week variety, short days should be started in the middle of October. For Thanksgiving flowering, start short days about September 15. They are grown at 60°F (15°C).

Cyclamen. It takes 18 months from seed to flower. Plants should receive shade from April 15 to September 15 and should be kept as cool as possible. The shade is removed after September 15, and the plants are grown at 50°F (10°C).

Gardenia. The last pinch of the plant stems should be made in mid-July. Thereafter the plants should be grown at 60°F (15°C) and should receive a humid atmosphere.

Kalanchoe. This plant is started from seed sown during the spring or summer months. It takes from 6 to 9 months from sowing to flowering. The plants must be given short days from September 1 to October 20 in order to flower for Christmas.

Poinsettia. All rooted cuttings and older plants must be lighted from 10 P.M. until 2 A.M. up to October 1 to 10, depending upon the variety grown.

Plants then are to receive short days and should begin to show color in their bracts around November 20 to 25.

Azalea. These plants are grown at 45–48°F (7–9°C) until 6 weeks before Easter. Any new growth that is visible should be pinched out to ensure flowers of the proper size.

Calceolaria. This plant is started from seed sown in August in a terrarium atmosphere. The seedling is transplanted into a 2¼-in. (5.6-cm) pot when two pairs of leaves have formed and is again repotted in 5–6 weeks into a 3-in. (7.5-cm) pot. In early November, the plant is repotted again into its final container for spring flowering.

Chrysanthemum. Follow the directions for chrysanthemum under "Christmas."

Cinneria. Sow seed in August and September in the same manner as for calceolaria. When seedlings reach one inch (3.5 cm) in height, pot in 2½-in. (5.6-cm) pots. They must be repotted into 4-in. (10-cm) pots a month later. In November or December, finally repot into 6–8-in. (15–20-cm) pots for spring flowering. The plants must be kept cool, 45–50°F (7–10°C), and care must be exercised not to overwater them directly after each repotting.

Daffodil. The bulbs must be stored for 10 weeks at 45–50°F (7–10°C). Once the plants have some top growth showing and good root growth, they can be taken out of storage and forced at 60°F (15°C). The forcing time for February flowering is 3–4 weeks, for March flowering 2½–3 weeks, and for April flowering approximately 2–2½ weeks.

Easter lily. The forcing time for precooled bulbs is 13 to 14 weeks at 60°F (15°C).

Gardenia. This plant should receive its last pinch in mid-August and then should be grown at 60°F (15°C) during the fall and winter months.

Hyacinth. These plants must be stored during the months of October and November at 40–45°F (4–7°C). Once the nose of the plant is 1½ in. (3.75 cm) high, they can be forced at 60–65°F (15–18°C) for February or early March flowering. For late March or early April flowering, the plants must be held in storage at 50°F (10°C) until two weeks before Easter and then may be forced at 60°F (15°C) in the greenhouse.

Hydrangea. For the formation of flower buds, the plants must receive a night temperature of 60°F (15°C) for 6 weeks. They are then placed in cold, dark storage at 35°F (2°C) for an additional 6 weeks (mid-October to December). This time is necessary for the development of the flower bud.

During this storage period, *leaf drop* must also occur; it is best accomplished by placing apples in the storage area and allowing them to give off ethylene gas. After the storage period, it takes 3 to 3½ months at 60°F (15°C) for flowering. For Mother's Day, keep the plants in storage until early February.

Pink flowers. Use sodium nitrate (nitrate of soda) and a fertilizer high in phosphorus and keep the soil pH above 6.2.

Blue flowers. Use ammonium sulfate and aluminum sulfate (or aluminum nitrate) and make certain that the soil pH stays between 4.5 and 5.0. No phosphorus should be added with the fertilizer.

Tulip. Potted bulbs are stored at 45–50°F (7–10°C) for a specific number of weeks determined by the variety grown. Forcing can be done once the tips or sprouts are just showing. For February and March flowering, the forcing time is 3 to 4 weeks at 60°F (15°C). For March and April flowering, the forcing time is 3 weeks. If the bulbs are not potted when placed in storage, add 2 to 3 weeks to the forcing times and make certain that the plants have developed a good root system before they are forced.

GREENHOUSE CONSTRUCTION

With the anticipation of additional leisure time, more gardeners are contemplating the erection of a greenhouse on their property. The more handy gardeners will decide to build their own, either from raw materials or from a prefabricated or precut model, while less handy individuals will have their greenhouse erected by professionals.

In either case, there are a variety of items that the gardener must consider before proceeding with the construction of any home greenhouse. They are

Site or location. A location should be selected that fits the property and its plantings, and which receives a minimum of three hours of winter sunlight daily. It should be convenient to the home and source of utilities, and should be positioned to capitalize on the best exposure to light, wind, and weather conditions. Ease of construction, installation, and yearly maintenance must all be considered when the location is chosen.

Style or type. The *free-standing* type provides the most growing room for the investment, but may be undesirable because of its isolation from the house, especially in bad weather. The *lean-to* style enables one wall of the house to be used as the fourth wall of the greenhouse. To minimize the loss of growing space within this style, the door should be in one of the end sections. The *attached* style combines the advantages of both the free-standing and the lean-to styles, because the narrow end that is joined to the home also contains a doorway, enabling the gardener to move from the home to the greenhouse rapidly. Some gardeners prefer the *glass-to-ground* style of greenhouse, because it allows for light beneath the benches, thus increasing the available growing space. The *regular* style is positioned on base walls and is easier to heat than the glass-to-ground type. It also provides a storage area beneath the benches.

Construction material.   Many hobby greenhouses are constructed from wood, with redwood, cedar, or Douglas fir being the most commonly used. There is less heat loss, and less expansion and contraction with wood than with aluminum. The use of aluminum may also detract from the home and its surroundings. However, one advantage that aluminum has over wood is that it is essentially maintenance-free.

Covering material.   *Glass* is the most economical, the longest lasting, most attractive covering; it also transmits the most light. Where glass breakage may be a problem, *fiberglass* is used. It is more expensive than glass and is rejected by some because it filters out some of the available sunlight, and is preferred by others for its heat retention ability. *Polyethylene plastic* is the cheapest of the three coverings, but it has the shortest life expectancy and gives the poorest light transmission.

Foundation.   The perimeter walls of the greenhouse must:

1. Go below the frost line, usually 3 ft (0.91 m) deep.
2. Extend at least 4 in. (10 cm) above the ground level.
3. Be 4–6 in. (10–15 cm) wide on the surface.
4. Be 8–32 in. (20–80 cm) high above the ground for regular style greenhouses.

Floor and walkways.   The best floor for a greenhouse is one of soil covered with gravel, while the walkways between the benches are usually made of concrete.

Installation.   In addition to the foundation, it is important to anticipate and plan for all those other steps that must be taken prior to the erection of the greenhouse. For example, should the siding be removed from the house area where the lean-to will be positioned? *Prefab* models are composed of individual pieces that must be assembled in proper sequence and positioned in the correct place. *Precut or preassembled* models are usually shipped in larger sections, so construction time is likely to be less.

Ventilation.   Some form of ventilation must be used in all greenhouses to provide plants with an exchange of air. The two most common types for hobby greenhouses are the *top opening*, which can be operated manually or electrically and the *electric* fan type, which forces the air out of the greenhouse.

Humidity.   As mentioned earlier, a greenhouse should be humid. This is accomplished manually by simply wetting down the floor of the greenhouse. Other methods of increasing the humidity level are spraying, misting, and the use of electric humidifiers.

Heating. This must be a prime concern of all greenhouse operators. Both the initial installation and the cost of long-term operation must be weighed before the heating unit is selected. Electric heaters are usually less expensive, are small in size, and do not require venting. Oil and gas, either natural or bottled, may be more economical over the long run, but both need venting.

There are two types of heat loss that must be considered when the heating requirements for any greenhouse are determined. They are (1) the transmission loss through the covering material, and (2) the loss of heat due to air infiltration.

Heat loss due to transmission is calculated by:

$$\text{Heat loss} = UA(t_i - t_o)$$

where

U = the heat transmission coefficient of the covering material (glass is 1.13 BTU/hour for each degree Fahrenheit difference).

A = the area of the greenhouse in square feet.

t_i = inside temperature (°F).

t_o = outside temperature (°F).

Heat loss due to air infiltration is calculated by:

$$\text{Heat loss} = 0.036V(t_i - t_o)$$

where

0.036 is the amount of heat in BTU's needed to raise one cubic foot of air 1°F (assume two complete air changes per hour), and V is the volume of the greenhouse in cubic feet.

Necessary conversions

$$1 \text{ ft}^3 = 0.028 \text{ m}^3$$

$$1°F = 0.44°C$$

$$1 \text{ ft}^2 = 0.092 \text{ m}^2$$

Chapter Twenty–Four

Floral Arrangements

For all florists, the arranging of flowers into centerpieces, corsages, wedding bouquets, and funeral sprays is a daily, year-around business.

When called upon, florists must blend their artistic talents with a knowledge of basic floral design. They are expected to know and make proper use of the available flowers and transform them into a lasting display of beauty.

The intention of this chapter is to expose the reader to all the necessary factors and techniques that are involved to become successful in the art of flower arranging.

COLOR
HARMONY

Color harmony in any floral arrangement may be obtained through the correct use of flowers, supporting greens, and the proper container. A complete understanding of the color wheel (Fig. 24-1) and how to use it will give any designer a good deal of flexibility in the creation of the final arrangement. The wheel contains a total of twelve colors of which three are known as primary colors (red, yellow, and blue), three are known as secondary colors (orange, green, and violet), and six as intermediate colors. Red and yellow are known as warm colors, while blue is considered a cool color.

Complementary colors are pairs of colors exactly opposite each other on the wheel, and these can be used to good advantage in any arrangement whenever the background is neutral. Analogous colors are those colors next to or close to each other on the wheel; when used in equal intensity, they can make a pleasant arrangement. It is important to emphasize to the beginner that dark colors must be kept low in any arrangement.

CONTAINERS

Besides the flowers, the correct container is the most important ingredient of any arrangement. The container selected must be deep enough to guarantee a continuous supply of water to the flowers and large enough to adequately hold and display the flower stems.

The container should also be simple in color and design, so that it does not detract from the arrangement.

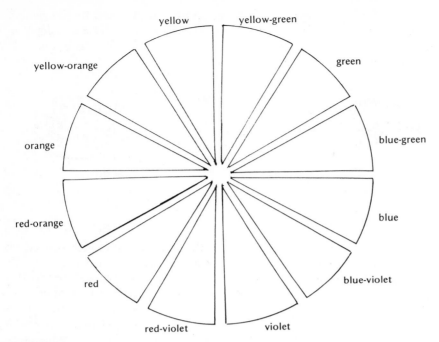

FIGURE 24-1
Color wheel.

The beginning floral arranger should start with four basic containers (Fig. 24-2), two of them having a height of 1½–2 in. (3.75–5 cm) and the other two being 4–6 in. (10–15 cm) tall. One of the containers of each height should be circular, and the other should be rectangular in shape.

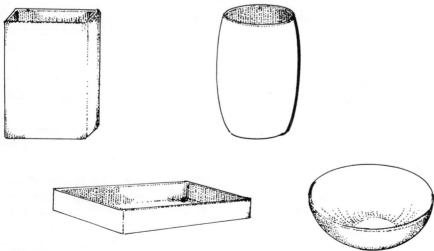

FIGURE 24-2
Four basic containers for floral arrangements.

HOLDING
MATERIALS

The holding material (Fig. 24-3) is necessary to keep the flowers in position within the arrangement. Shallow containers require a needlepoint holder held in place with florist's clay. When such a holder is employed, the flower stems must be cut straight across.

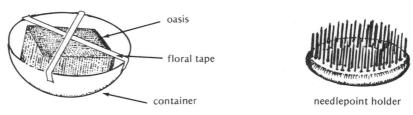

FIGURE 24-3
Holding materials for arrangements.

For deeper containers, crumpled chicken wire, shredded styrofoam, or oasis can be used. The oasis is a spongelike material that can be easily cut to fit the shape of the container and then can be soaked with water. It is advisable to keep all three of these materials in place with floral tape. Usually, two pieces are necessary; they should be positioned to make an X across the holder, but not in the exact center of the container, because it may interfere with the positioning of the central flower in the arrangement.

PRINCIPLES
OF FLORAL
DESIGN

The principles of design that guide a florist's creations are the following:

1. The natural habitat of the flowers selected must be known and used to good advantage. Flowers that grow in an upright manner naturally should be placed similarly in the arrangement.
2. The eventual design must be formulated in the florist's mind before he begins. To do this properly, the florist must first consult with the purchaser with respect to both price and the purpose intended for the arrangement.
3. Selection of the proper container is very important. It should be large enough to hold the flowers, but not so large that it distracts from the arrangement.
4. The addition of greens is the next step. The purpose of the greens is to hide the material that is used as the holder for the flowers and to add both body and contrast to the arrangement.
5. The colors used, coupled with the size, shape, and texture of the materials chosen, must be selected with care to achieve the proper balance, so that all parts of the arrangement will be in harmony with each other.
6. The flowers must not cross in the arrangement but should follow one of three lines: (1) vertical, (2) horizontal, (3) spiral, or (4) a combination of all three.

7. A floral arrangement may be either symmetrical or asymmetrical.

8. Each arrangement should have a center of interest or focal point. This ensures that the viewer will appreciate the arrangement as a complete, well-coordinated design and will not be distracted by one or several improperly positioned flowers.

9. All necessary steps should be taken to ensure that the arrangement remains fresh as long as possible. There are a wide variety of commercial fresheners used by florists in their arrangements. Their selection is based entirely on the type of arrangement requested.

*BASIC FLORAL
DESIGNS*

There are many different types of floral arrangements. The shape finally selected will be determined by the purpose intended and by where the arrangement will be displayed. These same factors will also affect the selection of the container.

In general, a tall container should contain flowers that are one and one-half to two times the height of the vase; in most low containers, the height of the tallest flower stems should not exceed one and one-half to two times the length or diameter of the bowl. All the flower stems should give the impression of originating from a central point.

Once the basic shape and container have been decided upon and the holding material has been covered with the appropriate greens, the tallest flowers are positioned first to define the outer boundaries of the arrangement. These are frequently referred to as *line flowers*. The form and filler flowers are then added to the arrangement to give it body and to fill in the empty spaces.

Occasionally, vivid or accent flowers are added for the purpose of enriching the arrangement.

Each arrangement must have a focal point, and for most types it is usually kept low and in the center of the arrangement. For simplicity, Table 24-1 presents the basic shapes for floral arrangements, plus a description of the most appropriate container type to use for each, a schematic diagram, and an explanation of where the focal point should be found.

Table 24-1
BASIC FLORAL ARRANGEMENTS

Name	Container type	Focal point	Diagram
Triangle (Pyramid) (Equilateral triangle)	low dish or bowl	center	
Side triangle (Right triangle)	flat tray or dish	at base	
Circular	circular bowl	lower half of circle	

Table 24-1(continued)
BASIC FLORAL ARRANGEMENTS

Name	Container type	Focal point	Diagram
Vertical (Perpendicular)	slender tall vase	center	
Horizontal	low container	center of arrangement when viewed from above	
Half circle (Convex)	flat or pedestal container	center, just above rim of container	
Oval	urn vase or tall round vase	center of flower mass	
S or Hogarth curve	cylindrical vase	center of S	
Crescent (Half moon)	low container	not well defined but necessary	

Chapter Twenty-Five

Conditioning Removed Plant Material

Pruning and propagation both involve removal of plant parts for their own specific reasons. In addition to these two practices, plant material may be removed to force woody stems into flower or to preserve flowers and foliage for dried arrangements.

The removal of the plant material will not by itself guarantee the desired result. It is important for the interested individual to know when the material should be removed from the parent plant, and how the material should be conditioned or treated prior to its proposed use.

For simplicity, the balance of this chapter will be presented in outline form, stating the correct procedures for obtaining and conditioning the plant material for the purpose intended and providing any additional information that will help to insure success.

Stems of certain woody plants may be collected from the parent plant, brought indoors, and forced into flower weeks before they would flower naturally outdoors. The procedure for doing so is the following: *FORCING WOODY STEMS*

1. Select branches 2–3 ft (0.61–0.91 m) long from the south side of the plant, four to six weeks before the plant normally flowers outdoors.
2. Select stems that contain many plump, round flower buds from the newest growth.
3. Cut each stem on a slant on a warm sunny day.
4. Place the stems in deep *warm* water for at least 30 minutes.
5. Submerge the stems in deep cold water and place them in a cool location for three to four days out of the direct sunlight.
6. Change the water every three to four days, making certain to recut each stem on a slant.
7. Keep the stems in the coolest living area of the home, because the duration of flowering time is dependent upon the temperature received by the stems, and the cooler the location, the longer the stems will remain in bloom.

Cut branches that may be forced

flowering cherry (Prunus)
flowering crabapple (Malus)
dogwood (Cornus florida)
forsythia (Forsythia)
Japanese quince (Chaenomeles)
Juneberry (Amelanchier)

magnolia (Magnolia soulangiana
  or stellata)
redbud (Cercis eanadensis)
Thunberg spirea (Spirea)
viburnum (Viburnum)
witch hazel (Hamamelis)

Shrubs that can be forced when grown in pots or tubs

azalea (Rhododendron)
daphne (Daphne)
deutzia (Deutzia)
Japanese quince (Chaenomeles)
mock orange (Philadelphus)

plum (Prunus)
rhododendron (Rhododendron)
spirea (Spirea)
spicebush (Benzoin aestivale)

Dried floral arrangements have grown in popularity during the past few years. This is due to their long-lasting qualities and to the numerous ways in which dried flowers can be used. The expense of commercially treated flowers and foliage may be prohibitive for many interested individuals. However, there are presently drying materials available that can be considered inexpensive, since these materials may be reused time after time.

The procedure for using silica gel for the drying of flowers, especially large-petaled ones, is as follows:

1. Select blossoms that are completely open during the early afternoon of a hot, clear day so that their moisture content is at its lowest level.
2. Cut each flower stem directly below the calyx before placing in the gel.
3. Use any airtight, nonmetallic box to house the gel.
4. Cover the bottom of the container completely with the gel before adding any flowers.
5. Place each flower, face up, in the gel and then gently sprinkle additional gel into the container to completely cover each flower and fill the container.
6. Cover the container and seal it with masking tape. Thin-petaled flowers will take three days, whereas the heavier blooms should remain in the gel for five to seven days.
7. After the required drying time has elapsed, pour off enough gel to partially expose the flowers and then gently lift each bloom from below with your fingers.
8. Remove any excess gel surrounding the flower with an artist's camel-hair brush.

1. The gel has a definite effect on certain flower colors.
   a. Blues become brighter.
   b. Yellows and oranges keep their color.
   c. Whites turn creamy or pinkish.
   d. Certain reds and pinks do not hold their color and change to maroon or lavender.
2. Fleshy flowers, such as gloxinia, petunia, and primrose, do not cure well in the gel.
3. Lilies, tulips, geraniums, and anemones must be immediately sprayed with clear plastic as soon as they are removed from the gel.
4. To recondition the gel, it should be spread out in a shallow metal pan and heated in an oven set at 150°F (65°C) until the crystals return to their original blue color.

Additional Information about Silica Gel

Flowers may also be dried by using a mixture of yellow corn meal and borax. Both ingredients are inexpensive and are easily purchased at the supermarket. The procedure for using this mixture is

1. The two ingredients are mixed in the ratio of one pound (0.45 kg) of yellow corn meal to 5 oz (140 g) of borax.
2. Use the same techniques mentioned under silica gel for cutting, positioning, and covering the flowers in any airtight nonmetallic container.
3. The drying time required is twice that of the gel. Thin-petaled flowers cured in three days in the gel will take at least a week to dry in this mixture.
4. The mixture can be dried after each use by placing it in an oven set at 150°F (65°C). During the drying time, the mixture must be frequently stirred and developing clumps must be crushed.

Natural air drying of flowers has been practiced for many years and is still relied upon as the best way to dry certain flowers. The procedure involved is the following:

1. Cut the branches when there is absolutely no moisture on any of the upper foliage and flowers. This is a necessary precaution against mildew.
2. Strip the lower foliage from each stem.
3. Tie a bunch of stems together and hang upside down in any out-of-the-way area that is dry and which has good air circulation.
4. Most flowers should be sufficiently dry and ready to use after three to four weeks.

Most plant foliage is now dried by using a glycerine-water mixture. The procedure for employing this mixture is the following:

1. Cut the foliage on a clear, hot day.
2. Remove all the leaves that will be below the level of the solution.
3. Use any container that is tall and slender, yet sufficiently broad at the base to support the foliage.
4. Use either of *two* commonly used mixtures. One contains one part glycerine to two parts boiling water; the other contains equal parts of glycerine and boiling water.
5. Plunge the prepared stems into the hot mixture and make daily observations of the change in color.
6. Once the top leaves show a change in color, the stems can be removed from the container. The process may take from five days to two weeks to completely change green foliage to dark brown color.
7. The final solution should be discarded and not reused, because most of the glycerine has been absorbed by the foliage.

Foliage that is available all year from the florist is boxwood, cedar, eucalyptus, fern, leather leaf, laurel, and rhododenron. This should be treated as soon as it is purchased.

From most home gardens, the following foliage is available: all evergreens, rhododenron, Japanese holly, laurel, leucothoe, and fern.

Many fresh floral arrangements can be created with flowers from your garden. The following procedure describes how to obtain them and what steps must be taken to properly treat a variety of flower stems to ensure their freshness.

1. Garden flowers should be cut either early in the morning or late in the day, provided that they are free of surface moisture.
2. The stems must be cut on a slant with a sharp knife or garden shears.
3. Before placing the flower stems in warm water, it may be necessary to treat the stem in one of the following ways:
   a. *Most annuals with soft stems.* Make a one-inch (2.5-cm) cut upwards through the center of the stem.
   b. *Woody perennials and shrubs.* Crush the bottom 2 in. (5 cm) of stem with a hammer.
   c. *Bulbous stems.* The base of the stem must be squeezed to remove the juice to prevent clogging and poor water uptake.
   d. *Hollow stems.* The stem tip must be either dipped into boiling water or singed with a flame.
4. Any excess foliage on the stems and all those leaves that will be below the waterline must be removed.
5. The stems should be placed in warm water, 110°F (43°C) for 2–3

hours before they are used in an arrangement. Enclosing them in plastic or covering them with paper helps to reduce the water loss.

Herbs grown for the purposes of flavoring and fragrance must be harvested properly. The required treatment is related to the plant part selected.

1. Foliage should be removed from the plant once the flowers start to open, since the herb oils are the most abundant at this time.
2. Flowers should be cut when in full bloom.
3. The roots of herbs must be harvested in the autumn after the top growth has ceased.
4. When herb seeds are desired, the seed head must be completely dry before the seeds can be extracted and used.

Chapter Twenty-Six
The Compost Pile

Most home gardeners construct a compost pile as the means to maintaining the organic content of their soils. An additional benefit is the personal satisfaction derived from the fact that stockpiling plant wastes, table scraps, sawdust, and other formerly living materials helps to solve the problem of solid waste disposal.

Compost is used to enrich all types of garden soils and as a top dressing on existing lawns. Whenever it is to be used indoors, either as a potting soil or blended with other materials for seed germination, it must be pasteurized. This is best accomplished by placing the compost in an outdoor enclosed grill and exposing it to a temperature of 200°F (93°C) for at least 30 minutes.

REQUIREMENTS FOR SUCCESSFUL COMPOSTING

Composting involves the partial decomposition of organic material by the action of bacteria and other microorganisms. To achieve the greatest success in the shortest possible time, there are definite requirements that must be met:

1. Proper aeration. The microbes must receive an adequate supply of air throughout the entire pile, or decomposition will be greatly reduced. Turning the pile once a month during the warmer months helps to overcome poor aeration, and at the same time enables the excess CO_2 to be released from the pile.

2. Sufficient moisture. The ideal moisture content of any compost pile should range between 50 and 70 percent. Whenever the water content of the pile is excessive, nutrients are leached from the pile. A high moisture content can be decreased by turning the pile. The top of the pile should be depressed and should be lower than the sides to help the pile receive and absorb rainfall, thereby keeping the moisture content high. During periods of drought, the pile must be watered.

3. Nutrients. Some source of nitrogen must be added to the pile to replenish that removed by the microorganisms and to help maintain the bulkiness of the pile. Potash and phosphate are usually not required, even though most experts agree that a complete fertilizer should be added to every other layer of the pile during its construction [one cup for every 30 ft² (2.76 m²)].

4. Temperature. The rate of decomposition is very slow at temperatures near freezing, but as the temperature rises, the rate doubles for every increase of 18°F (10°C) provided that there is sufficient oxygen for the microbes.

5. Source of microbes. They are present on all plant material and are of two types. Microbes of the first type grow and decompose plant material at ordinary temperatures and are responsible for the initial decomposition. They are, however, destroyed when temperatures within the pile reach 115°F (43°C). Fortunately, microbes of the second type are able to continue the decomposition and are responsible for raising the temperature within the pile to 150–170°F (66–77°C).

6. Addition of soil. This is not always necessary, but it helps to conserve nutrients, retain moisture, and supply additional microorganisms.

7. Proper location. The pile must be placed in a partially shaded, well-drained, level location that has a reasonably solid foundation. If placed in a sunny spot, the intense summer's heat may kill many of the microbes. The minimum size recommended for a compost pile is 4 ft (1.2 m) wide, 3 ft (0.91 m) high, and any length desired. It can be left in a loose pile or can be enclosed within a structure of wire, wooden boards, or snow fencing.

8. Ground limestone. This is added to each layer not receiving fertilizer and is done to ensure that the pile does not become too acidic, which could destroy the action of the microorganisms.

It is important for the gardener with extremely limited space to know that the entire composting operation can be performed in plastic leaf bags. Leaves, fertilizer, soil and moisture are placed within the bag after being completely mixed. The bag is then sealed and placed in any appropriate storage place for the decomposition to occur.

Chapter Twenty–Seven

Organic Gardening

Organic gardening is a specific method of growing vegetables without the use of commercial pesticides or inorganic fertilizers. It emphasizes the recycling of a variety of waste materials either by composting or by direct addition of these materials to the soil. As mentioned earlier in Chapter 2, organic matter is invaluable as an addition to any soil, no matter whether it is applied as manure, as a cover crop, or as a soil-improving crop.

The fertility of the soil is maintained by a complex interaction between the available organic matter and the soil microorganisms, plus the periodic applications of lime, to raise the soil pH, and organic fertilizers. In early spring, organic fertilizers are not very effective, because the soil temperatures necessary for their breakdown by the soil microorganisms are too low.

It must be understood by everyone that even though organic gardening can be extremely successful for most home gardeners, it would be completely impossible to employ its principles on a wide commercial scale to feed the population of this country.

SOIL-IMPROVING CROPS   Soil-improving crops, also known as green-manure crops, should be turned under before they become mature, because then their rate of decay is much faster than when they become thoroughly dry. These crops are grown during the same growing season as vegetables for the purpose of preparing the soil for the growth of future crops grown in succeeding years.

Cover crops are usually grown during the dormant season when vegetable crops are not grown for the combined purpose of soil improvement and soil protection.

Both the soil-improvement and soil-cover crops should be turned under before they become mature and at a time when the soil is both warm and well supplied with moisture.

To help the inexperienced gardener achieve some degree of success with an organically grown vegetable garden, the following suggestions or remedies are offered:

1. Select disease-resistant varieties.
2. Use the following "organic" materials for insect and disease control:
   a. Dormant oil spray for aphid, scale, and mites.
   b. Nicotine sulfate for aphid and pear psylla.
   c. Rotenone for aphid.
   d. Ryania for aphid, codling moth, and Japanese beetle.
   e. Wettable sulfur for diseases, such as scab and brown rot.
3. Use beneficial insects, such as ladybugs and praying mantises, to devour other insects.
4. Use certain "companion plants" to rid certain plants of specific pests, for example:

Companion plant	Desired plant	Removes
Garlic	Roses	Blackspot, mildew, aphids
Parsley	Roses	Rose beetles
Beans	Potatoes	Mexican bean beetle, potato beetle
Soybeans	Corn	Chinch bugs, Japanese beetles
Chives	Peach	Peach borer
Radishes	All cucurbits	Striped or spotted cucumber beetle
Chives and garlic	Lettuce, peas	Aphids
Herbs	Cabbage	Cabbage butterflies

5. Use certain "plant traps" so that insects attack the trap, and leave the more valuable plant material alone, for example:

Plant trap	Valuable plant	To lure
Marigolds	Tomato, bean, eggplant	Nematodes
Marigolds	Cucurbits	Cucumber beetle
Sunflower	Corn	Most corn insects
Sage	Broccoli	Cabbage moth
Geraniums	Rose, grape	Jap. beetles
Marigolds	Corn	Jap. beetle
Mint or Rosemary	Cabbage	Cabbage moth
Horseradish	Potato	Colorado potato beetle
Nasturtium	Bean	Mexican bean beetle
	broccoli	Aphids
	cucurbits	Cucumber beetle

Chapter Twenty–Eight

House Plants

At some time during his life, almost everyone has either purchased or received some type of house plant. Sometimes a house plant is selected because of a special day or holiday and at other times simply to fulfill a desire to grow and care for a living plant.

Certain house plants are noted for their foliage, others for their attractive flowers, some for their fruit, and still others for both their flowers and their colorful foliage.

Like all living things, house plants have scientific or generic names. Some are sold by their genus or scientific name, while others are sold by a variety of common names. Because of this, three people can refer to the same house plant by three different names. To add to the confusion, many house plants bearing the same name can have one variety with variegated leaves while another variety has plain green leaves.

To grow a specific house plant, its owner should become familiar with the cultural requirements of that plant, such as its temperature, light, watering, and fertilization requirements. Certain plants are extremely specific in their growth requirements, while others will survive the most adverse conditions.

TYPES OF HOUSE PLANTS

House plants are generally divided into the following five groups:

1. *Flowering plants.* These usually need daily light to produce the desirable flowers.
2. *Foliage plants.* These plants are able to exist without much available sunlight.
3. *Fruit-bearing plants.* They need abundant light, warm temperatures, and sufficient water to produce their fruits.
4. *Succulents.* These plants are capable of living at very high temperatures; they need very little water.
5. *Vines.* Some vines are grown for their flowers, others for their foliage, and still others are selected because they can be trained to grow on foreign objects.

New house plants can be obtained from seeds, cuttings, bulbs, tubers, leaves, and from the roots. Certain plants have plant patents, which restrict completely any propagation without payment to their originators.

The balance of this chapter will be an alphabetical presentation of the more commonly grown house plants. Important information will be given with each house plant to enable the reader to achieve a much greater degree of success in growing a specific plant.

African violets. These colorful flowering plants are probably today one of the most popular house plants. They bloom continually when grown properly. They can be grown under artificial fluorescent lights; they should have from 12 to 18 hours of light daily. The genus name for these plants is *Saintpaulia*. They are not related to true violets.

Aluminum plant or *Pilea cadieri* is a neat bushy plant that is well covered with silver-striped and spotted leaves. It provides a nice contrast with other green plants, because its leaves are either green or bronze.

Aralia is a tall, airy plant with ribbonlike leaves which should be kept warm and should receive plenty of water. It should not be mistaken for a fern.

Azaleas are primarily grown in a greenhouse and forced into flower for a specific event or holiday. They are very decorative when in bloom. They can be forced into bloom in the home by placing the plant outdoors during the summer months and then bringing it inside around Labor Day and providing it with ample heat and light.

Baby tears is a creeping plant that forms a dense mat of tiny green leaves. It is very hardy and extremely easy to propagate. Its generic name is *Helxine* and its low-growing habit makes it suitable for terrariums.

Begonias are one of the most popular plant families grown indoors. Most begonias are grown for their flowers, although the rex begonia is grown primarily for its foliage. The flowering types are propagated from stem cuttings, while the rex begonia is propagated from leaf cuttings.

Bouganvillea can withstand a variety of growing conditions but prefers a soil on the dry side. This plant will bloom all winter, if pruned to maintain a bushy, pot-size plant.

Bromeliads are native American plants whose flowers and foliage make them desirable as house plants. They adapt readily to all indoor conditions. The pineapple is a member of this family.

Browallia displays quantities of violet-blue or white starry flowers on trailing stems. It can be grown from seed or purchased as a small plant. It is not specific about its growing conditions and is best displayed in hanging baskets.

Burn plant or *Aloe vera* is a small succulent with heavy leaves originating from a central base. The leaves may be plain green or variegated. Its common name is associated with the fact that the gelatinous pulp oozing from a removed leaf has analgesic properties for the treatment of simple burns.

Caladium is grown from a tuber and is planted in a shaded area. All varieties are variegated and display at least two colors in each leaf. Most caladium leaves are 12–15 in. (30–37.5 cm) long.

Camellia is a house plant requiring very specific growing conditions: plenty of room for growth, an average temperature about 50°F (10°C), and a very humid atmosphere. These conditions are usually obtained by placing the plant in an enclosed porch or sunroom. If necessary, it must be pruned and repotted directly after it blooms.

Chinese evergreen is an inexpensive, slow-growing foliage plant that tolerates hot, dry rooms. These plants will remain green even when neglected in dark corners and dim light. The genus name is *Aglaonema*.

Christmas cactus or *Zygocactus* is commonly referred to as crab's claw. It flowers freely in December and January when kept dry from mid-September until November except for a light sprinkling every ten days and when kept in the dark during the evenings, because it responds to short-day conditions.

Chrysanthemums flower naturally during periods of short days. They now can be programmed to flower on a specific date. Chrysanthemums come in a variety of sizes, shapes, and colors, and can be grown as either cut flowers or as potted plants.

Coleus thrives best when grown in full sun. It can be easily raised from seed or from stem cuttings. The leaves of this plant usually contain a blend of two or more colors, and the flower is small and insignificant.

Crossandra is a flowering plant that requires temperatures in the seventies. The blossoms appear at the top of the plant. It is a difficult plant to flower in the home.

Croton is a colorful foliage plant that is taller than most grown. These plants require a warm temperature, abundant light, and sufficient moisture.

Crown of thorns or *Euphorbia splendens* is a succulent plant that thrives in a hot, dry environment. This plant's chief danger is overwatering. It is not a cactus. The branches of this plant can be trained on wire, and its flowers form constantly on thorny stems.

Cyclamen is an attractive plant of the primrose family that requires 18 months of growing time from seed to flower. The plants are grown in a very cool greenhouse to flower for the Christmas holiday season.

Dieffenbachia is also called dumb cane, because a piece of the leaf placed on a person's tongue is supposed to render him temporarily speechless for about 24 hours. All species of this genus tolerate poor growing conditions.

Dracenas are a varied group of foliage plants referred to as corn plants, because of their characteristically shaped leaves. These plants will live forever in high temperatures with plenty of water.

English ivy or *Hedera helix* is a favorite vine grown in the home. Its leaves may be green or variegated and varied in shape, depending upon the variety selected.

Episcia or *flame violet* has trumpet-shaped flowers with exquisite and interesting foliage. It is best displayed in a hanging basket. It is related to the African violet and gloxinia.

Ferns are popular as house plants. The most tolerant choices are rabbit's foot, holly fern, bird's nest, and European hart's tongue. All ferns should be soaked at least once a week and kept out of direct sunlight. The staghorn fern is an easy fern to grow but requires the right location within the home.

Fittonia is sometimes called nerve or mosaic plant. This plant displays its broad, strongly veined leaves on trailing stems. They are easier to grow in a greenhouse than in the home. They are frequently planted in hanging baskets.

Flamingo flower or *Anthurium* prefers high humidity, but resents transplanting. Varieties of this plant offer long-lasting flowers in a variety of colors.

Fuchsias are the favorite of many for their hanging baskets. They are not true house plants and will become dormant during the fall months after flowering has occurred.

Gardenia is a long-lived house plant that produces strongly fragrant white flowers. Numerous flower buds develop, but they require very specific growing conditions in order to bloom, high humidity being the most important. Their leaves should be a glossy, bright green at all times, and if they become light green or chlorotic, an application of chelated iron usually corrects this problem.

Geranium is a very common house plant that does much better outdoors than it does inside the house. During the winter months, most plants become elongated, lose many of their leaves and produce few flowers, but even then, most homeowners do not discard them. There is a wide range of differing varieties of geranium, which is increasing each year with the new hybrid varieties.

German ivy or *Senecio mikanioides* is a vine that grows well in a hanging basket. Its appeal is due to its large, dark green leaves, fragrant yellow flowers, and its ability to thrive over a wide temperature range.

Gloxinias are easy to grow from tubers. They may also be propagated from leaf cuttings. When kept warm with a minimum temperature of 60°F (15°C) and watered with care, these plants will flower when they are ready. When their growth ceases, the tubers are stored in a frost-free place until the following February or March.

Grape ivy is one of the most popular and decorative foliage vines for the home. The leaves are glossy and are three-parted. It can be allowed to trail or be trained to grow on bark. The genus name is *Cissus*; it is not related to a true ivy.

Hoya is a flowering vine that is commonly called *wax plant*. The plants grow slowly and will not flower until they reach a certain size or age. It is sometimes planted in hanging baskets.

Impatiens is grown from seed, producing a variety of colors. These plants come as close to continuously flowering as any house plant grown.

Their most important growth requirement is sufficient water, and new plants are easily obtained from the parent plant by taking stem cuttings.

Jade is a member of the Crassula family characterized by its crown of branches, which bear small, glossy succulent green leaves. It has been a favorite house plant for a long time for those people desiring a nondemanding one.

Jasmine is a fragrant flowering plant that is easily grown indoors if it receives sufficient sunlight, a temperature about 65°F (19°C), and regular watering. The name jasmine is a general name that encompasses a wide variety of house plants.

Jerusalem cherry is the least expensive, most common, and heaviest-bearing fruit plant that bears its fruit in the month of December. The fruit, however, is not good to eat. New plants can be quickly obtained from the seeds of a ripe cherry taken during the month of February.

Kalanchoe is a succulent plant prized for its clusters of small scarlet or white blossoms. These plants flower when exposed to short-day conditions during the fall. They are usually started from seed in the summer or early fall to flower some 5 to 6 months later for Christmas or Valentine's Day. Additional plants can be obtained by taking stem cuttings from the mature plant.

Lemon is a small citrus tree whose fragrant, waxy, white flowers appear at the same time as the fruit does.

Lime is another citrus tree that has bright green fruits and fragrant blossoms all year long.

Monstera is sometimes called Philodendron pertusum, Swiss cheese plant, or split leaf philodendron. The fruit, if it ever forms, would taste like a cross between a banana and a pineapple.

Orange is another citrus plant with pink-tinted blossoms that develop into plum-sized orange fruit.

Orchids, once thought to be impossible to grow by the amateur grower, are becoming more popular. There are many types and varieties that can be successfully grown by the homeowner. The growing conditions vary somewhat from one variety to another, but they all desire a more humid atmosphere than the average home can provide. This problem is normally solved by placing pans of water near the plants.

Oxalis is an old-fashioned bulbous plant with masses of shamrock-like leaves with small yellow, lavender, pink, or white blossoms appearing as early as August and continuing throughout the winter.

Palms are durable, slow-growing, tolerant of high temperatures and insufficient light, and require little care, except for a regular schedule of watering.

Peperomia is a small, bushy, foliage plant having red stems and silver-striped leaves suggesting a watermelon. They can be grown warm, but should not be overwatered. The leaves on these plants may be plain green or variegated.

Philodendrons are the easiest of all the vines to grow. They grow well even in very dim light. The most popular species has shiny red, pointed buds that uncurl into its characteristic heart-shaped leaves.

Pilea is an easily grown house plant having glossy, dark green, flat leaves with interesting veins. They have small clusters of flowers that appear to originate from the leaves. Another variety of pilea, aluminum plant, has already been mentioned.

Poinsettia is a Christmas flowering plant with either red, white, or pink bracts. The real flowers are small and insignificant at the terminal growing tip. Poinsettias respond to short days and may not flower if the duration of light is too long. They will flower in the home if given complete darkness from 5 P.M. until 7 A.M. each night for approximately 40 days or 6 weeks, starting about October 1.

Pothos is sometimes called *devil's ivy*. It is an excellent vine for warm rooms, displaying thick, oval, green leaves with splashes of yellow. The genus name is *Scindapsus*. Pothos can be trained to grow on bark.

Prayer plant derives it name from its flat, broad, oval leaves that fold together vertically at night, like a person's hands held in prayer. The genus name for this plant is *Maranta*. The roots of another species of Maranta are a source of tapioca.

Purple passion or velvet plant has fleshy green leaves covered with purple hairs, giving the leaf a velvet appearance. It prefers a warm room and moderate light. Its scientific name is *Gynura aurantiaca*.

Rheo discolor is also called *Moses on a raft*. It is an easily grown house plant with leaves that are red or purple on the underside and green on top. All the leaves originate from a central stalk, while the small white flowers originate near the base of the leaves.

Rubber plant is a very rugged house plant which, even when neglected, thrives in dark corners of the home. When it is exposed to sunlight, the new leaves are red in color. The genus name for this plant is *Ficus*, meaning fig. Many varieties are grown as house plants, ranging from the upright types to the creeping types.

Sansevieria or *snake plant* is a member of the lily family that may eventually produce green flowers. It is one of the toughest house plants, surviving both poor growing conditions and amazing abuse. New plants are easily propagated by placing a one-inch leaf section in sand.

Sensitive plant is a unique house plant whose leaflets collapse when touched. New plants are started from seed. Its genus is *Mimosa*, a member of the legume family.

Shrimp plant or *Beloperone guttata* is a plant with pinkish bracts that surround its white flowers. It requires plenty of sunlight and abundant water, both in the pot and on the foliage.

Spider plant or *Chlorophytum elatum* has narrow green leaves, striped with white, that grow in small tufts. Additional plants can be quickly obtained by removing the tufts that hang over the side of the pot and potting them. This plant does yield small white flowers, which are normally hidden by the mass of foliage.

Strawberry geranium is often called *mother of a thousand* because of its tendency to produce long runners at whose tips new plants form. These plants may form delicate white flowers. The genus name is *Saxifraga*, and plants can be purchased with plain green or variegated leaves.

Tradescantia or *wandering jew* is a slow-growing vine, with brittle stems and colored leaves. It is easily propagated from stem cuttings. Numerous varieties are now available for house culture.

Zebra plant is a reliable house plant desirable for both its flowers and foliage as long as the humidity level is relatively high. The flowers are pale yellow, while the leaves are dark green with white veins. The genus name is *Aphelandra.*

Chapter Twenty–Nine

Lighting Plants Indoors

Experimentation has shown that the light requirement for normal plant growth is that of the visible spectrum plus the far-red rays. Red light in the visible range encourages flowering, while the blue rays promote foliage growth. The far-red rays are responsible for environmental changes, such as fall color and cessation of growth in the fall.

To achieve optimum growth of plants indoors under artificial light, it is necessary to employ "balanced light." This requires the use of both fluorescent and incandescent bulbs in the ratio of 3 watts of fluorescent light to every watt of incandescent light.

BALANCED LIGHTING

The fluorescent tube recommended is the cool-white, which has a minimum wattage of 40. This type supplies both red and blue rays to the plant. The incandescent bulb selected is best used with a wide reflector that guarantees that the light emitted is cast downward. The incandescent bulb is needed for the all-important far-red rays. It also emits red rays.

Since the incandescent bulb does produce far-red or heat rays, it is imperative to provide the plants with the ventilation necessary to remove the excessive heat.

The lights should be positioned at a height of 12–18 in. (30–45 cm) above most plants, from the tube to the rim of the pot. The normal duration of light required by plants varies from 12 to 18 hours. As a general rule, the closer the light source to the plants, the shorter the time that plants must be lighted; i.e., plants grown 12 in. (30 cm) from the lights should be lighted for 12 hours.

Experimentation has shown that certain plants are extremely specific in their requirements for light intensity or brightness. Two 40-watt fluorescent tubes positioned 12 in. (30 cm) above the plants will yield an average of 400 foot-candles of light intensity over the area served by these two tubes. The intensity is greatest directly under the tubes and decreases in any direction from the center.

One method of determining the light intensity delivered by an indoor setup is to use a photographic light meter. Set the meter at an ASA rating of 25. Next, adjust for a lens opening of f/5.6. Hold the light meter 12 in. (30 cm) above the plants and read as follows:

Meter's indicated shutter speed	Approximate foot-candles
2	45
1	80
$1/2$	110
$1/4$	150
$1/8$	260
$1/15$	590
$1/30$	1000
$1/60$	2000
$1/125$	4000

This procedure can also be used in any home or office to learn the existing light conditions. Once a determination has been made, proper selection of plant material to fit the desired location can be accomplished with a high degree of success.

Plants that grow at light levels between 50 and 250 foot-candles are

Baby's tears (*Helxine soleiroli*)
Cast iron plant (*Aspidistra elatior*)
Century plant (*Agave americana*)
Gold dust plant (*Aucuba japonica*)
Moses in a boat (*Rheo discolor*)
Pepper plant (*Piper nigrum*)

Between 250 and 650 foot-candles the following plants will grow, prosper, and flower:

African violets (*Saintpaulia*)
Amaryllis
Crown of thorns (*Euphorbia splendens*)
German ivy (*Senecio mikanioides*)
Gloxinia (*Sinningia*)
Oxalis
Shrimp plant (*Beleperone guttata*)

Those plants that thrive at 650 to 1400 foot-candles are

Begonias
Calceolaria
Christmas cherry (*Solanum pseudocapsicum*)
Christmas pepper (*Capsicum frutescens*)
Coleus
Flame violets (*Episcias*)
Kalanchoe
Primrose
Seeds (both vegetable and flower)

The following flower seeds require light for germination:

African violet   Kalanchoe
Ageratum   Lobelia
Alyssum   Pansy
Browallia   Petunia
Calceolaria   Primula
Cineraria   Salvia
Coleus   Snapdragon
Gloxinia   Streptocarpus
Impatiens   Violet

These seeds require darkness for germination:

Calendula   Phlox
Cyclamen   Statice
Dahlia   Verbena
Larkspur   Zinnia

For the proper construction and arrangement of lights for indoor plants, the following suggestions should be taken into consideration:

1. It is better to use several incandescent bulbs of low wattage than one large one to reduce the buildup of heat produced by these bulbs. They should be positioned between the rows of fluorescent tubes and inserted into porcelain-coated reflectors. These reflectors are unnecessary when incandescent bulbs with built-in reflectors are used.

2. The narrow boxlike fixtures designed to hold a single fluorescent tube may be used, because they can be mounted close together. Once in position, this type would require some type of reflector to direct the light downward towards the plants.

3. When four fluorescent tubes are used, the incandescent bulbs should be positioned between the second and third bank of fluorescent tubes (Fig. 29-1).

4. When the fluorescent tubes are to be placed within an enclosed growth chamber, the chamber should be at least one foot longer than the length of the tubes used. This will make tube replacement much easier.

5. The longer the fluorescent tube used, the greater will be its intensity. The production of more flowers should result.

6. When the growing area is completely enclosed, the plants must be provided with adequate ventilation so that they will receive the change of air they need.

7. A shield made of glass, plastic, or Plexiglas positioned 2 in. (5 cm) below the incandescent bulbs keeps the heat away from the plants.

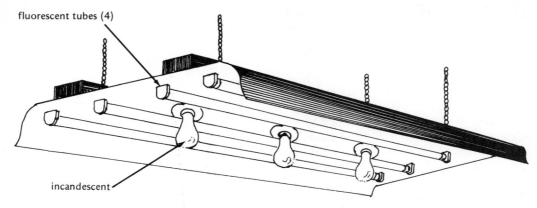

fluorescent tubes (4)

incandescent

FIGURE 29-1
Arrangement of lights for indoor growing.

The slight loss in light intensity due to the shield's presence is not as damaging to the plants as the excessive heat would be without the shield.

Chapter Thirty

The Forcing of Bulbs in the Home

In this chapter, bulbs (tubers and corms) will be separated into three distinct groups for forcing procedures. The first group discussed is the *tender* bulbs, which require *no* cold-temperature treatment and can be planted and placed directly in the home.

The second group of bulbs treated is classified as *semi-hardy*. These bulbs must be placed in a cold frame or unheated garage for 6 to 8 weeks before they can be brought into the house and forced.

The *hardy* bulbs are the third group to be discussed. These must receive a temperature of 40°F (4°C) for 8 to 12 weeks before they can be forced inside the home. They can be placed in an unheated garage, a cold frame, or buried outdoors.

Prior to the forcing of any bulb, tuber, or corm, the gardener should be familiar with the following general information about bulbs:

BULB INFORMATION

1. Only top-quality bulbs should be selected. They should be firm and clean.
2. Any container selected for forcing must have a hole for drainage, and the soil mixture used must provide the bulb with good drainage.
3. The depth of the container used for forcing must be at least twice that of the bulb for proper root growth.
4. Small bulbs are planted just below the soil's surface, while larger bulbs are usually not completely covered.
5. All bulbs must be thoroughly watered as soon as they are planted and must be kept moist during the storage period.
6. Before any bulb can be brought into the house and be forced into flower, its root system must be well developed, with roots visible through the drainage hole.
7. During the low-temperature storage period, the bulbs must be kept in the dark. This is accomplished by either burying the pot outdoors or placing another pot on top of the one containing the bulbs.

TENDER BULBS

Achimenes are planted in February or March to flower from May to October. They are frequently planted three to five to a basket, in a 5-in. (12.5-cm) hanging basket and forced at a temperature of 60°F (15°C). They should be stored dry until their growth starts again in late winter.

Amaryllis or *Hippeastrum* can be planted one to a 6-in. (15-cm) pot from December to June to flower 4 to 5 weeks later. They may be planted outside after the danger of frost has passed, but they must be repotted before the first fall frost and allowed to dry out and rest before they can be forced again.

Tuberous begonias are planted in an equal mixture of peat moss and sand during the months of January through March. Make certain to plant the tuber with the depression side up. After flowering has ceased, dry the tuber completely and store in sand at a temperature of 45–60°F (7–15°C) until the following January.

Caladium tubers are planted in January to produce their bi-colored leaves during the summer months. One tuber is planted in a 6-in. (15-cm) pot. They should be kept out of direct sunlight, but they can take temperatures in the eighties. In September, they must be dried off and stored dry until January.

Clivia or *Kafir lily* is planted in the spring, kept at a temperature between 60 and 70°F (15 and 21°C) during the summer, and given a rest period in the fall at a temperature of 50°F (10°C), to ensure flowering the following year during March and April.

Colchicum or *meadow-saffron* needs no pot or soil to flower in the home during September and October. If planted in the garden before the frost, it will flower each fall.

Convallaria or *lily of the valley* may be forced throughout the year from pips set in watertight containers. They will flower about three weeks after being planted and should be discarded after blooming. Make certain to maintain the water level in the container so that the pips do not dry out.

Cyclamen corms are planted during the months of August through October to flower about four months later. They should be kept in a cold location where the temperature remains between 45 and 55°F (7 and 13°C). After flowering is over, the corm is kept in the pot and placed in a cool basement; it must be kept moist until the next August.

Gloxinia or *Sinningia* tubers are planted, one per 5-in. (12.5-cm) pot, in a mixture of sand and peat moss during February to June. These plants will flower for a long period of time, but when they have finished flowering, they should be stored in a cool basement and not be allowed to become dry. Forcing temperatures can range from 60–70°F (15–21°C).

Iris bulbs are planted from August to November in a variety of containers, the most important requirement being that there be at least ½ in. (1.25 cm) between bulbs. Tall varieties are Dutch iris, while the short varieties are *Iris reticulata*. These bulbs will flower about five months after planting. Once flowering has passed and the foliage has been allowed to dry completely, the bulbs are removed from their container and stored dry until the following August.

182

Lycoris or *golden spider lily* bulbs are planted in June, one per 6-in. (15-cm) pot, and kept at a temperature of 50–55°F (10–13°C). Its normal flowering time ranges from November to January. The grower of this bulb must be familiar with the fact that it goes dormant during the spring and summer months.

Paper white daffodils or *Narcissus* are planted in any watertight container from October to March. The level of the water in the container must be constantly just below the base of the bulb. Before forcing can occur, roots must be visible through the drainage hole. The planting medium recommended is fine gravel rather than soil. The bulbs must be kept in the dark during the period necessary for root formation, approximately one week. The bulbs can be forced at temperatures from 60–80°F (15–27°C), and when flowering is completed the bulbs should be discarded.

Oxalis bulbs or tubers are planted in September, several to a 5-in. (12.5-cm) pot to flower later throughout the year. After flowering has finished, the bulbs should be stored dry in sand until the following September or October.

Smithiantha or *Naegelia* tubers are planted during February and March for summer flowering. One tuber is placed in a 6-in. (15-cm) pot and forced at a temperature of 60–70°F (15–21°C). The treatment after flowering is the same as for gloxinias.

Zantedeschia bulbs can be planted at two distinct times of year, in the fall to flower during the months of January to April, or in the spring to flower during the summer months. A 6-in. (15-cm) pot will house one bulb, which should be forced at temperatures between 65 and 70°F (18–24°C). The bulbs are stored dry after flowering has occurred until the next planting period.

Several *Allium* bulbs are planted in a pot during September and October to flower usually during March. As soon as there is evidence of roots growing through the drainage hole, the pots can be forced at a temperature between 45 and 50°F (7 and 10°C).

Anemone or *windflower* is planted in September for an eventual flowering during February through April. The tubers must be soaked in water prior to planting and are forced into bloom at temperatures between 40 and 50°F (4 and 10°C). After flowering, the tubers must be stored in a cool shady area until September.

Freesia corms are planted during August, 12 in an 8-in. (20-cm) pot. When given temperatures between 45 and 55°F (7 and 13°C), they will flower from December until March. The foliage should be allowed to dry naturally before the corms are dried off and stored in dry sand until the following August.

Ixia or *corn lily* is planted in the same manner as freesia. The bulbs are potted in September or October and will flower during February and March when forced at 45–55°F (7–13°C). Treatment after flowering is similar to the freesias.

SEMI-HARDY BULBS

Lilium or *Easter lily* requires approximately 115 days to flower when given a precooling treatment. For normal forcing without precooling, the bulbs are usually planted in August or September, one per 6-in. (15-cm) pot, and grown at forcing temperatures of 60°F (15°C) to flower at Easter. Once flowering has passed, the bulb may be planted in the garden.

Ornithogalum or *sea onion* has small bulbs that are planted from September until December to flower three months later. They are forced into flower at temperatures between 55 and 65°F (13 and 18°C) and are planted six to a 6-in. (15-cm) pot.

Ranunculus or *Persian buttercup* tubers are planted six to a 6-in. (15-cm) pot in August and September to flower during February, March, and April. They are forced into flower and treated after flowering in the same manner as anemones.

Tritonia or *Montbietia* bulbs are planted in August and September, six to a 6-in. (15-cm) pot. They are forced at temperatures between 50 and 55°F (10 and 13°C) and are treated in the same manner as ixia after flowering has ended.

HARDY BULBS

Chionodoxa or *glory of the snow* bulbs are small, and one 6-in. (15-cm) pot can house from six to twelve bulbs, depending upon the grower's preference. They are planted in September and October, forced at a temperature of 60°F (15°C), and should flower in February and March.

Crocus corms are planted in August and September, several to one 4-in. (10-cm) pot. They should be forced at 50–55°F (10–13°C); they burst into bloom in February and March.

Eranthis or *winter aconite* is planted in September in the same manner as crocus, and is forced at 55°F (13°C) to flower in January and February. After flowering has passed, the bulbs are removed from their container and planted outdoors.

Erythronium or *dog's tooth violet* is treated just like the crocus. Its bulbs also can be planted outdoors after they have flowered.

Fritillaria or *guinea hen flower* should be treated in the same manner suggested for Eranthis.

Galanthus or *snowdrop* is treated just like crocus, except that the forcing temperature should range from 45–50°F (7–10°C).

Hyacinth bulbs are usually planted three to four per 5-in. (12.5-cm) pot in September. After storage, they are forced at 60°F (15°C) and should flower between January and March. After the foliage has completely died down, the bulbs can be planted outside.

Ipheion or *Triteleia* or *spring starflower* bulbs are planted four to a 6-in. (15-cm) pot in September, are forced at 45–55°F (7–13°C) and should flower in February and March.

Leucojum or *spring snowflake* bulbs are planted several in a 6-in. (15-cm) pot during September and October. They are forced at 45–50°F (7–10°C) and should flower sometime from January to March. Like all hardy bulbs, these also can be planted in the ground outdoors after being forced inside the home.

Muscari or *grape hyacinth* are treated just like Leucojum, except that the forcing temperature should be between 50 and 65°F (10 and 19°C).

Narcissus or *daffodil* bulbs are planted in September and are closely packed in their container, with one bulb touching the next one. When forced at 50–65°F (10–19°C), they will flower from January to April.

Scilla bulbs are small and are, therefore, planted several to each 6-in. (15-cm) pot in September and October. They are forced and should flower as the Narcissus mentioned above. The two most popular varieties grown are Siberian squill and wild hyacinth.

Tulipa or *tulip* bulbs are planted six to eight per 6-in. (15-cm) pot in September or October, are forced at 60–70°F (15–21°C) and should flower sometime between December and April. They also can be planted outside after flowering indoors.

In addition to the suggestions previously covered about bulbs, the following information will aid the gardener whose only interest is in outdoor planting:

NATURAL GROWTH OF BULBS OUTDOORS

1. Each bulb must be planted at its recommended depth. The general rule associated with this suggestion is: The larger the bulb, the deeper it must be placed below the soil's surface. Small bulbs, like crocus, are usually planted 2–3 in. (5–7.5 cm) below the ground.

2. Contact with manure will destroy most bulbs.

3. Do *not* remove the foliage after flowering. Allow it to die and dry up naturally. This practice ensures the storage of the necessary food for the following year's bloom.

4. The ideal time to dig up crowded bulbs is in June. They can be separated and immediately replanted, or they can be stored in a cool, dry place until planting time.

5. The order of bloom for outdoor planting is presented below:

Table 30-1
ORDER OF BLOOM FOR OUTDOOR PLANTING

Early spring	Late spring	Summer	Fall
Galanthus	Narcissus	Allium	Colchicum
Leucojum	Tulip	Camassia	Crocus
Eranthis	Hyacinth	Iris	Sternbergia
Chionodoxa	Scilla	Leucojum aestivum— summer snowflake	
Scilla		Lily	
Puschkinia		Lycoris	
Muscari			
Crocus			
Fritillaria			

Chapter Thirty-One

Terrariums

The forerunners of the popular terrariums of today were known as Wardian cases. They were so named in honor of a London surgeon, Dr. Nathaniel Ward, whose most noteworthy accomplishment was the construction of large glass containers designed to house plants while they were shipped around the world. After being subjected to long and difficult journeys aboard ship, these plants not only survived the trip but usually arrived in excellent condition.

In principle, the terrarium should operate like a self-contained water cycle. The moisture in the soil is initially absorbed by the roots, transported throughout the plant to the leaves, transpired through the leaves, condensed on the inner walls of the container, and eventually returned to the soil. The air within a terrarium is always near its saturation or dew point.

Most terrariums provide humidity levels higher than 75 percent, enabling plants that could not be grown in the normal home atmosphere to actually thrive within the confines of its walls.

The choice of glass containers that can be used for terrariums is endless and ranges from aquariums, candy jars, Victorian domes, and bottles to the newer bubble-type bowls. It is important that, whatever the container selected, it be clear glass and not be colored or tinted, and that each container have a top or cover.

SOIL MIXTURE

The soil mixture most commonly recommended for use in terrariums contains one part loam, one part humus, and a third equal part of sand or perlite. Granular charcoal must be added to the soil mixture to keep the soil from becoming too acidic.

Some of the containers selected for terrarium growth may have a provision for drainage, but most do not. In either case, a layer of coarse gravel or small stones, at least an inch (2.5 cm) thick, is placed on the bottom of the container. On top of the gravel should be placed an inch (2.5 cm) layer of sand (Fig. 31-1). Some growers prefer to spread a thin layer of coarse sphagnum moss over the sand before adding the soil mixture to the container. For a more pleasant appearance, it is suggested that the soil in the

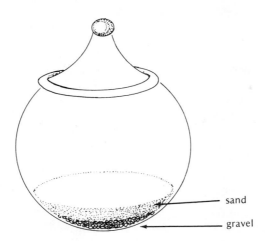

FIGURE 31-1
Drainage for terrariums.

sand

gravel

container be arranged on a slight sloping contour. This makes it possible for the viewer to appreciate all the plants in the container at the same instant.

The soil should be slightly damp, but not soaking wet before the plants are positioned. It is advisable to decide upon your plant arrangement before planting begins. Once you have done so, place the plants one at a time in the container, digging a hole just large enough to accommodate the plant's earth ball. After all the plants are properly positioned, the soil must be carefully watered by adding very small quantities of water intermittently over the next two hours until the soil surface appears moist. This procedure must be done with extreme care so that the entire contents do not become thoroughly drenched. Those containers without a provision for drainage require an even more careful procedure.

After the watering procedure has been completed, it is advisable to wipe the inside walls of the container with a soft tissue. This is done to remove any excess water and soil that has been spilled on the walls.

Once the plants are properly positioned, the taller ones in the rear of the container and the smaller ones in the foreground, and they are properly watered, the container should not require additional moisture for several months unless it is left open.

If any terrarium container clouds up with moisture, this can easily be corrected by opening the cover for a short period of time. Some growers suggest removal of the top each evening for a few hours.

The terrarium should never be placed in direct sunlight. Its preference is for a well-lighted setting, such as an east- or west-facing window. If the only possible location available is a south-facing window, the container will require some type of shading from the sun.

There are a wide range of plants that are ideal for terrarium growth. For simplicity, they will be presented in two separate lists, cultivated plants and plants found growing wild.

Cultivated Plants

Small orchids
Ferns
Fittonia
Baby's tears
Sinningia pusilla
Small palms
Pellionia
Begonias
Succulents

Small ivies
Salaginella
Pilea
Maranta
Small crotons
Peperomia
Aluminum plant
Saintpaulia
Sedum

Strawberry geranium
Philodendron
Wandering Jew
Boxwood
Plectranthus
Parlor palm
Alternanthera
Cactus
Jade

Wild Plants

Partridge berry
Wintergreen
Ground pine
Moss

Lichens
Evergreen seedlings
Small wild ferns
Club moss

Pipsissewa
Hepatica
Rattlesnake plantain

Chapter Thirty–Two

Ferns Grown as House Plants

For any degree of success to be achieved with ferns within the home, their culture must be understood first. They should be given plenty of light but not placed in direct sunlight. The growing medium must be light and porous and ideally should contain two parts loam, two parts leaf mold, and one part sand or perlite.

They prefer a temperature above 55°F (13°C) and should be kept away from radiators and drafts. Ferns are normally watered by immersing the entire pot in a tub of water and allowing the excess to drain off. They should be provided with ample moisture both in the soil and the surrounding air. When the humidity level becomes very low, fern leaves or fronds may turn yellow.

Each fern should be given sufficient growing room, because when crowded, its leaves or fronds may become damaged. They are normally repotted only once a year and fertilized only during periods of active growth. Any runners that form should be treated as suckers or excess growth and should be cut off as soon as they appear, since they use food that is needed by the plant.

Because some of the ferns presented in the ensuing paragraphs may have several common names, they will be listed in alphabetical order according to their scientific names.

Adiantum cuneatum (maidenhair fern) has black shining leaf stems and fan-shaped leaflets. It is best grown in a terrarium, since it has difficulty growing in most homes because they are too dry.

Asplenium nidus (bird's nest fern) is a durable and popular fern for the home. It grows too large to be used in terrariums. The arrangement of its fronds gives the viewer an understanding of its common name.

Asplenium bulbiferum or *A. viviparum* (mother spleenwort ferns) has leaves that resemble carrot leaves. The small bulblets that form on the fronds will yield new plants once they become mature. The new plant, when removed, must contain a small piece of the parent. This piece must be planted just below the surface of the pot to ensure success.

Blechnum brasiliense (fern tree) is a very sturdy fern with a trunk that can grow to 3 ft (0.91 m) in height and fronds that may be 2 ft (0.61 m) long.

Crytomium falcatum (holly fern) has leaflets with a shape, texture, and contour similar to holly leaves. Each new frond that develops on this fern has one more pair of leaves than the previous frond.

Davallia fejeensis (rabbit's foot fern) is characterized by having hairy, brown, creeping rhizomes that will eventually grow over the side of the pot. Copper wire, in a hairpin design, is used to fasten the rhizomes to the surface of the soil when this fern is repotted. This fern is well displayed in a hanging basket.

Davallia bullata is commonly called squirrel's foot fern.

Jacaranda pedata is a small fern suitable for terrariums and dish gardens.

Nephrolepsis exaltata or Boston fern and its varieties, ruffle leaf and sword fern, are all fast-growing ferns that lend themselves to display in hanging baskets.

Pellaea rotundifolia (button fern) is a member of a large group of ferns having both evergreen and deciduous leaves. The fronds grow to 2 ft (0.61 m) in height and have round or spearlike leaflets. This fern can grow in low levels of light intensity and is popular in hanging baskets.

Phyllitis scolopendrium cristatum (hart's tongue fern) has foliage that resembles a tongue. This fern must be kept below 70°F (21°C) and receive good air circulation.

Platycerium alicorne (staghorn fern) is an epiphitic fern grown usually in osmunda fiber on slabs of wood. When mature, it resembles a stag's head, with its divided fronds. This fern must be kept in an atmosphere of high humidity and must be frequently misted to prevent it from becoming dry.

Polypodium aureum is another scientific name for rabbit's foot fern, squirrel's tail fern and bear's paw fern. The correct generic name is Davallia.

Polystichum adiantiforme (leatherleaf fern) is able to exist under extremely low light intensities.

Polystichum tsus-simense is a miniature fern with dark green leathery fronds. It usually does not grow over 8 in. (20 cm) high and is, therefore, popular in terrariums.

Pteris cretica and P. serrulate are commonly called brake or table ferns. All varieties of Pteris have varied and attractive ribbonlike fronds that remain small in size. Some of the varieties may be difficult to grow properly in warm, dry homes. They can be used in the larger-sized terrariums. Some varieties of Pteris are variegated.

Some plants that are commonly referred to as ferns are not true ferns. They are:

Asparagus ferns. All the varieties of asparagus fern belong to the lily family; specific ones are used by florists in their arrangements and corsages.

Pilea microphylla or *artillery fern.* This plant receives its common name from its greenish flowers, which discharge a cloud of pollen when shaken. It is a small plant suitable for dish gardens and terrariums.

Chapter Thirty–Three

Common Vines Grown in the Home

Many plant admirers who have difficulty growing flowering plants indoors often select a vine in an attempt to satisfy their desire to grow a living green plant. There are vines available that can withstand the most adverse indoor conditions.

The general culture associated with most vines is that they should be planted in a well-drained potting soil, should be watered when they become dry, and should be grown within a temperature range of 60–75°F (15–22°C). Most vines are well displayed and thrive in hanging baskets. They prefer medium humidity levels and are all propagated by cuttings. The few vines that are propagated by other methods will be mentioned in their individual presentations, which follow.

Allamanda is a sun-loving vine that has yellow or purple flowers in the summer. It should be cut back and repotted in the month of January.

Clerodendrum is a fast-growing vine that produces white, red, or violet flowers during the spring and summer months. It does extremely well in full sunlight.

Creeping fig (*Ficus pumila*) has delicate, small, heart-shaped green leaves. Its aerial roots will cling to any object. It is used in hanging baskets and in dish gardens as a ground cover. It prefers high levels of humidity, and its light requirements can range from indirect to full sun.

Devil's ivy or *Pothos* (*Scindapsus aureus*) can exist in any light conditions and humidity levels but is unable to tolerate cold air or water. It is a vine characterized by green oval leaves splashed with yellow or white.

Varieties of *English ivy* (*Hedera helix*) are extremely easy to grow and can be trained to climb or can be allowed to trail. The leaves of the different varieties have various shapes, and some may be variegated. They should be kept well pruned. Hahn's maple ivy is one of the best varieties to be grown indoors because of its dark green, waxy, overlapping leaves and its ability to withstand low light intensities.

German ivy (*Senecio mikanioides*) has fragrant yellow flowers and deep green leaves. It prefers to be grown cooler than most vines and to be kept more moist than most.

Grape ivy (*Cissus rhombifolia*) is recognized by its glossy three-parted leaves. It is a slow-growing, sturdy vine that can do well in low light.

Kangaroo vine (*Cissus antarctica*) is a rapid-growing vine that is free of pests. It is identified by its long, narrow, shiny, heart-shaped leaves with ragged edges. It can be grown under practically any light conditions and can be propagated by division.

Kenilworth ivy (*Cymbalaria muralis*) has small violet flowers tinged with yellow. Its kidney-shaped leaves are tinted underneath with red. This vine must be pruned and misted on a regular basis. Another common name for this vine is "mother of thousands." It can be propagated from seed.

Maurandya (*M. erubescens*) is a sun-loving vine that has arrow-shaped leaves and tubular flowers that form during the winter months. The color of the flowers ranges from blue to rose to white.

Philodendron (*P. cordatum*) is the easiest of all the vines to grow. Its shiny red buds uncurl into heart-shaped leaves. Under extremely low light intensities, the leaves on this vine become very small.

Pussy ears (*Cyanotis somaliensis*) is a relative of *Tradescantia*, having fleshy leaves covered with white hairs and trailing stems. It prefers to be grown in good light.

Ragwort vine or *"little pickles"* (*Othonna capensis*) flowers during the winter months, yielding small pickle-shaped fruit. It is best displayed in a hanging basket.

String of hearts vine or *"rosary vine"* (*Ceropegia woodii*) is recognized by its urn-shaped flowers and heart-shaped leaves that cascade downward from the edges of a hanging basket.

String of pearls (*Senecio rowleyanus*) has small flowers that appear during the winter time, giving rise to round fruits that resemble small green pearls. To flower, this vine must receive medium to high levels of light.

Swedish ivy (*Plectranthus australis*) is a popular vine that has either plain green or variegated leaves. It prefers to be grown in full sun and must be kept pinched to remain compact and neat in appearance.

Teddy-bear plant (*Cyanotis kewensis*) has trailing leaves and stems covered with brown hairs. It is related to *Tradescantia* and is usually grown in full sun. Stem cuttings should be frequently taken from this vine to keep it from becoming leggy.

Thunbergia (*T. alata* and *T. grandiflora*) is a vine grown for its large, showy, white (*T. alata*) or blue (*T. grandiflora*) flowers. These vines must be pruned in both fall and spring and must be misted regularly. They grow best when they receive direct sunlight and fresh air. Propagation can also be from seed.

Wandering Jew (*Tradescantia* or *Zebrina*) encompasses a large and varied family whose leaves usually are two-colored and whose stems are weak. All varieties must be constantly pinched back to prevent legginess. Flower color ranges from white to pink to purple.

Wax plant (*Hoya carnosa*) is another vine that can be grown at temperatures in the fifties (10–14°C). It is characterized by waxy leaves and clusters of small, light pink, fragrant flowers. New plants can be obtained by layering.

Chapter Thirty-Four
Fruit Plants Grown Indoors

Any avocado seed that has started to split open can be used to produce a desirable indoor foliage plant. Flowers may form but will never
yield a fruit indoors.

Avocado (Persea Americana)

Three toothpicks are inserted into the sides of the seed and are used
to support the seed while the base of the seed is immersed in a glass of
water.

After a few weeks in a warm, lighted location, roots and shoots should
appear. At this time the seed should be planted in soil and should be kept
trimmed to maintain a bushy plant.

Most citrus plants grown indoors for their fruits are best obtained by
grafting or from cuttings. Seeds can be used to produce new plants, but
flowers and fruit may never materialize.

Citrus

All indoor citrus plants must be watered once or twice a week by
soaking the entire pot, must be syringed to keep the foliage neat and
insect-free, and must be pruned in the spring or early summer just before
they are planted outdoors in a shaded spot, with the rim of the pot level
with the ground. This is recommended procedure for those desiring flowers
and eventual fruit during the fall and winter months indoors.

To ensure the formation of fruit on any indoor-grown citrus, it is
necessary that its owner hand-pollinate the flowers, taking the pollen from
one blossom and placing it on the stigma of another blossom.

The lemon grown indoors is *Citrus limonia ponderosa,* and the lime
is *Citrus aurantifolia.* There are several oranges grown indoors, of which
Citrus mitis and *Citrus nobilis deliciosa* are the two most common edible
ones. *Citrus taitensis* is an orange, producing small golf-ball-sized fruits
that are not edible.

Unroasted coffee beans are needed to produce this desirable hardy
indoor plant with its shiny, dark green leaves. They must first be soaked in
water for three to seven days to soften the outer seed coat before they can
be sown in any rooting medium. Germination and early seedling growth is

Coffee Tree (Coffea Arabica)

best accomplished at a temperature of 70°F (21°C). Thereafter, the plant should be grown at 60°F (15°C) and should be constantly pruned if flowers and beans are desired.

Date Palm
(Phoenix Dactylifera)

To start a plant from seed, unpasteurized dates must be purchased from a health food store. Several seeds should be sown per pot in a sandy soil. A temperature of at least 80°F (26°C) is needed for germination, which may take several months to occur and may yield only one or two seedlings.

Once germination does occur, pot the seedling in the normal potting mixture and place the palm in a warm, sunny window. This palm will not yield fruit, because the male and female flowers form on separate plants.

Indoor Fig
(Ficus Deltoidea)

This evergreen shrub is a popular house plant because of its slow growth rate and its ability to hold its fruit throughout most of the year, and because it is relatively pest-free.

New plants started from cuttings during the dormant winter season should be placed in sand and kept cool until spring. Once they form a callous, they are planted either in pots or in the garden. These plants may bear fruit during their second year of growth, provided that they are grown in a well-lighted location at a temperature of 60°F (15°C). The fruit will form in the axils of the leaves.

Jerusalem Cherry
(Solanum
Pseudo-capsicum)

In order for this plant to produce its inedible red fruit for Christmas, the seed from a mature fruit must be sown the previous January or February. This plant must be grown at a temperature of 50°F (10°C), must receive plenty of water, and must be kept away from drafts.

It is usually placed outdoors in May and kept well watered. During its flowering period, however, which normally occurs during August and September, it is advisable to decrease the quantity of water somewhat to help the plant set its fruit.

Upon the plant's return to the home, its stem tips should be pinched slightly to curtail further growth, and it should be placed in a well-lighted, cool, but moist location. *Solanum capsicastrum* is a close relative that has pointed instead of round fruits.

Kumquat
(Fortunella Japonica)

The propagation and culture for this freely fruiting plant is similar to that of the citrus. The small, round, orange-colored fruits against a background of green leaves make this plant a desirable ornamental house plant.

Ornamental Pepper
(Capsicum
Annuum or
C. Frutescens)

A variety of peppers can be grown as house plants from seed. They should be sown in a well-drained soil containing sufficient organic matter. Depending upon the variety selected, the fruits will vary in shape from slender, pointed, scarlet ones to round green or black fruits.

Pineapple
(Ananas Comosus)

The pineapple can be grown in the house as a foliage plant by rooting the top portion cut from a fruit. A small section of the fruit must be left attached to the leafy top, and several of the lower leaves must be removed.

Once this has been done, the top is allowed to air-dry for one to two days before it is placed in a shallow bowl containing pebbles or is potted in a sandy soil. The foliage should be kept moist by frequent misting without causing the base of the plant to become too wet. If this happens, the complete top will rot.

A pineapple plant can be forced to yield flowers that will eventually form fruit by enclosing the entire plant in a polyethylene bag containing three to four apples for four to five days. The apples produce ethylene gas, which speeds the respiration rate of the plant, resulting in the formation of both flowers and fruit. New leaves will become apparent soon after the plant is removed from the plastic bag. These will finally hold the desired fruit.

Pomegranate (Punica Granatum)

For this plant to finally produce its small, edible, orange fruit in the home, it must be propagated asexually from a stem section and not from seed, which do not breed true.

Indoor culture of the pomegranate requires a temperature of 70°F (21°C), good light at all times, and a continuously moist soil. It should be pruned whenever its stems become leggy. During the winter, it becomes partially dormant and sheds its leaves.

Sweet Potato (Ipomoea Batatas)

To form a sweet potato vine that makes a nice appearance in a hanging basket, select a tuber that is beginning to sprout. The tuber should be washed thoroughly to remove any chemical residue before being immersed in a container of water to half its length.

Toothpicks are positioned into the middle of the tuber and are used to support the potato while its roots are forming. The lower end of the tuber must be constantly covered with water.

Roots should form within two weeks, provided that the tuber is placed in a cool, dark location. Once they become apparent, give the sweet potato sufficient light so that the top growth will develop within an additional two weeks.

Chapter Thirty-Five

Bonsai

Bonsai is the art of training any tree or shrub to remain small or dwarfed in size through the use of specialized horticultural techniques. This miniature, old-looking plant will eventually be grown in a small suitable container and will be positioned within the container in a manner suggesting the plant's natural growth habit.

BASIC BONSAI TECHNIQUES   The basic techniques used to create bonsai include pruning, wiring branches, root pruning, and keeping the plant healthy. Pruning is used to systematically remove unwanted roots and branches in an effort to achieve the desired shape. Coiled copper wire is most commonly used to wire specific branches for the purpose of controlling both their shape and growth. The wire is left in place only until its presence might injure the bark.

Normally, root pruning is done each time the plant is repotted, which may be done every one to three years. This technique ensures that the size of the soil ball will continuously be confined and restricted to a predetermined diameter.

During the years required to achieve the desired bonsai effect, the plant must be watered when necessary. Fertilization is limited to one application per month of a complete liquid fertilizer. This technique guarantees continued but suppressed growth. Most bonsai require a time span of three to six years to attain the desired art form.

For your initial attempt at bonsai, select a young, vigorous, well-branched plant with a stocky trunk and a growth habit suggesting the potential of being transformed into an interesting form.

The container that will eventually house the plant must be chosen with care so that it harmonizes both in color and proportion with the plant. In most containers, the plant is positioned slightly off center. The small plastic caps found in the drainage holes of most bonsai containers should not be removed until planting is completed.

Many bonsai are created with a similarity to oriental floral arrangements, which involve a triangle, suggesting harmony between man, heaven, and earth (Fig. 35-1). The tallest or strongest main branch represents heaven, an intermediate-size branch represents man, and the lowest or third branch represents earth. For some bonsai, additional branches can

FIGURE 35-1
Bonsai—displaying the oriental triangle.

be used, provided that they are not distracting to the overall view and provided that they remain odd in total number and still include the three primary branches.

The soil mixture for most bonsai contains two parts loam, two parts perlite, and one part sand; it should be kept moist, but never wet, at all times as a precaution against the plant's becoming too dry.

There are certain house plants that are readily adaptable to bonsai, but the majority of plants used for bonsai are hardy species that are grown outdoors for most of the year. During severe winter conditions, they are usually brought into the house to be enjoyed and admired. Table 35-1 gives the reader a suggested list of hardy plants that can be used for bonsai. For beginners, it is advisable to start with either rooted cuttings or seedlings and to confine their height so that they do not exceed 2 ft (0.6 m).

Table 35-1
HARDY PLANTS SUITABLE FOR BONSAI

Eastern U.S.		South, Southwest, West Coast	
Azalea	Holly	Bamboo	Juniper
Barberry	Juniper	Camellia	Oak
Beech	Maple	Cedar	Olive
Birch	Oak	Citrus	Pineapple guava
Boxwood	Peach	Crape myrtle	Pittosporum
Cedar	Pine	Cycad	Pomegranate
Cotoneaster	Quince	Cypress	Spruce
Cypress	Redbud	Desert willow	Strawberry guava
Fir	Spruce	Douglar fir	Tamarisk
Firethorn	Sweet gum	Firethorn	Umbrella plant
Ginkgo	Willow	Japanese apricot	Western hemlock
Goldenrain tree	Wisteria		
Hawthorn	Yew		
Hemlock			

Chapter Thirty-Six

Cacti and Succulents

All cacti belong to the family known as Cactaceae. Most cacti have fleshy above-ground parts that are designed to conserve and store water in their tissues, which is also the identifying feature of succulents. Therefore, almost all cacti are succulents, although not all succulents are cacti, since they are members of other plant families.

Since both cacti and succulents are very similar in their cultural requirements, the general information concerning the correct culture of each will be presented together.

The soil mixture most commonly used is an equal mixture of a porous loam and sand. Sand taken from the beach must not be used because of its high salt content.

These plants should receive meager amounts of fertilizer only during their growing season, which for most occurs in the late spring and summer. Fertilizer must never be added during their winter rest period.

Usually in March, April, or May, just prior to the active growing season, each cactus and succulent should be knocked out of its pot to determine whether or not the plant should be repotted. If the roots have completely engulfed the container, the plant should be repotted to the next larger container. Fertilizer should not be added at this time, because the plant will have enough problems in adapting to its new environment.

The watering procedure recommended for these plants is to submerge the pot in a pan of water until the pot's surface appears moist. After this is accomplished, the soil is allowed to dry out until it is practically dry before it is watered again. During the winter rest period, even less water is required.

Cacti may be propagated from cuttings taken in the spring or early summer. Once cut, they should be left exposed to the air to dry for a few days before being placed in sand to root. Plants may also be obtained from seed. They should be sown in a sifted sandy soil in the spring and kept moist at a temperature of about 65°F (19°C).

Most succulents are easily propagated vegetatively from a variety of above-ground parts, such as leaves, stems, and suckers. Those succulents that form flowers can be propagated from seed.

A hobby enjoyed by some cactus growers is grafting one cactus onto another tall erect cactus, usually *Pereskia aculeata*. Cuttings of Pereskia are rooted in sand, and once they have reached the desired height, the grafting can be performed. A wedge-shaped piece is removed from the top of the Pereskia cutting, which is then ready to receive the V-shaped scion. The scion is placed on top and is held intact with a common pin. Thereafter, it is placed in any glass enclosed container until the union of the graft is complete. (See Laboratory Exercise 19.)

Cacti and the smaller succulents are sometimes used to fill a dish garden. The shallow container is first lined with gravel or small pebbles for drainage. A thin layer of the proper soil mixture is spread over the gravel. Next, the plants are placed in a suitable setting in the dish and additional soil containing charcoal is packed firmly around their roots. Sphagnum moss, sand, or small colored pebbles may be used to cover the soil and also to help conserve moisture. Once planted, the dish garden must be carefully watered so that the plants are not left waterlogged.

The variety of cacti and succulents grown as house plants is enormous. The following presentation offers the reader a listing of the more commonly grown plants:

Cacti

Christmas cactus	*Zygocactus truncatus*
Easter lily cactus	*Echinopsis*
Night blooming cereus	*Epiphyllum oxypetalum*
Orchid cactus	*Epiphyllum*
Peanut cactus	*Chamaecereus silvestri*

Succulents

Air plant	*Kalanchoe pinnata*
Bromeliads	*Aechmea*
	Billbergia
	Guzmania
Carrion flowers	*Stapelia*
Century plant	*Agave*
Crown of thorns	*Euphorbia splendens*
Hen and chicks	*Echeveria*
Jade	*Crassula*
Kalanchoe	*Kalanchoe blossfeldiana*
Living stones (pebble plants)	*Lithops*
Partridge breast	*Aloe* variegata
Sedum	*Sedum* species

Chapter Thirty–Seven

Plants for the Home and Garden

Five popular plant families will be discussed in this chapter: begonias, chrysanthemums, geraniums, gesneriads, and roses. The general cultural requirements will be presented for each family, plus any additional information associated with the various types or species grown.

BEGONIAS

As a family of house and garden plants, begonias are second only to African violets in popularity. The belief by many that begonias are difficult to grow indoors is rapidly changing to the new-found realization that all types are similar in their general culture, even though each type may appear to be different because of a dormant period and flowering time different from the others.

All begonias prefer a growing medium that is well drained and highly organic. They grow best in the temperature range of 60–65°F (15–19°C) and in a humid atmosphere, and they require protection from the summer's intense heat. When new growth becomes apparent on any begonia, it should be repotted into a larger container. Fertilizer, in liquid form, is recommended only during the period of active growth and never during the plant's dormant period. Begonias thrive best in locations where the air circulation is excellent; they must be shaded during the summer.

Their leaves should be sprayed with water at least once a week, and the plants themselves can be watered from above or from below. No matter whether they are grown indoors or out, they must be kept away from drafts.

Begonias are divided into five groups or classifications. Each of the five types will be presented separately with its characteristics and its special cultural requirements.

Fibrous Rooted

This group is represented by the wax begonia (*Begonia semperflorens*), which is noted for its small size and continuous profusion of red, white, and pink blooms throughout the year, when grown below 70°F (21°C).

These plants are propagated by seeds sown in the late winter for use as bedding plants in the garden, or from stem cuttings taken at any time of year. Cuttings taken during the spring and summer months root much faster and produce larger plants than those taken during the fall and winter, when both the intensity of light and its duration are decreasing.

To ensure a continuous display of flowers, the faded ones must be pinched off religiously. The wax begonia must receive long daylight conditions when grown at a temperature above 70°F (21°C), or flowers will not result.

During the winter months, the plant's round evergreen leaves may fall if it experiences a temperature below 55°F (13°C), a dry atmosphere, or a constant diet of being overfertilized and underwatered.

Semi-tuberous

This group of begonias consists of hybrids obtained by crossing *B. socotrana* and *B. dregel*. Plants in this group, also called Christmas begonia, are propagated by either leaf petiole cuttings or stem cuttings taken from the parent plant during the winter or early spring. When subjected to short-day conditions, varieties of this begonia type will produce either red, rose, pink, orange, or white flowers at Christmas time. Once formed, the flowers should last for two to three months.

Tuberous

Tuberous-rooted begonias may be propagated from seed sown during the winter months or from tubers planted in the early spring. This type of begonia responds to photoperiod, with long-day conditions causing flowering while short-day conditions result in tuber formation.

As soon as the buds on the tuber begin to swell, it can be planted. To keep it from rotting, the tuber must not be completely covered with soil. Tuberous begonias come in a variety of colors, and during their entire flowering period they should be fertilized each week.

Once the top growth has died down in the late fall, the tuber should be lifted from its container, allowed to dry, and then stored at a temperature of 50°F (10°C) until the following February or March.

Rhizomatous

Beefsteak begonia, or *B. feastii*, is the most frequently grown plant of this group. It is propagated by division of its rhizomes or by stem cuttings. Its pale pink flowers usually form on tall stalks that rise above its leaves and stems.

Its leaves are red underneath, green on top, and hairy near the edges, making it a desirable begonia for any flower garden.

Rex Group

Begonia rex and its hybrids are grown for their beauty and the coloring of their foliage, rather than for their flowers, since many of the varieties of rex begonia will never flower in the home. The leaves of all rex begonias are beautifully marked, hairy, and reasonably wide. They are most commonly propagated from leaf cuttings (see Chapter 7), although they may also be propagated by division of the plant; some of the newer hybrids are grown from seed.

During the summer months, rex begonias grow rapidly when placed in a warm, moist atmosphere with adequate light. To ensure continuous growth at this time, the plant needs ample quantities of both water and fertilizer.

From October to April, rex begonias should not be fertilized and should receive only enough water to keep the plant from drying out completely. During this period, it can receive plenty of light but must be kept in a warm, dry atmosphere.

Other Popular
Begonias

1. *Angel wing begonia* (*B. coccinea*). It has green wing-shaped leaves with white dots and is bronze underneath. Its flowers are salmon pink and the plant blooms all year. It is fibrous-rooted.
2. *Trout begonia* (*B. argenteoguttata*). Its leaves are oval, pointed, coarsely toothed, and olive green in color with white spots.
3. *Rieger begonias.* These are new hybrids that form mutations freely, but can not be propagated commercially unless a license and royalty fee is paid. They are excellent for hanging baskets and will flower continuously in either the home or greenhouse. Fertilization is best accomplished by using a weak fish emulsion.
4. *Iron cross begonia* (*B. masoniana*). This rex type has bright green, round leaves with brown markings.
5. *Begonia scharffi* has large pink clusters of flowers that bloom continuously. It has both dark green and red leaves. It is fibrous rooted.

CHRYSAN-
THEMUMS

Chrysanthemums belong to the composite family, whose members are characterized by having two types of florets or small flowers. The *ray* flowers are commonly referred to as petals and are found on the outside of the flower head, while the *disc* flowers are found in the center or eye of the flower. Most double flower types lack this type of floret.

The classification of chrysanthemums is based on two separate items: where they are grown and the type of bloom. Either they are grown in the greenhouse at a temperature of 60°F (15°C) and are usually not hardy when placed outdoor, or they can be grown outdoors in the garden where they must be hardy to withstand the severe winter weather.

Chrysanthemums are commonly divided into the following seven types of bloom:

1. Incurved, where all the florets curve inwards.
2. Reflexed, whose florets point outwards with the outermost ones even drooping.
3. Intermediate varieties, characterized by having florets that are partially incurved or semi-reflexed.
4. Pompons, which come in a variety of forms, most having spherical blooms.
5. Singles or "daisy" type, surrounded by ray florets.

6. Spoons, which have ray florets that are spoon-shaped.
7. Anemones, identified by having a cushionlike center surrounded by rows of ray florets.

Plants grown in the greenhouse flower after they have received the proper number of weeks of short days, a photoperiod response. From catalogues that offer chrysanthemum cuttings for sale, commercial growers select the varieties that will fit their needs and growing schedules with respect to the number of weeks of short days the plant must receive to flower, the color of the variety, and its type of bloom. The vegetative period required of all cuttings prior to the start of short days varies with the variety grown, the time of year, and the plant's eventual height.

The hardy outdoor chrysanthemums flower more as a response to temperature than to photoperiod, but both are involved. To obtain well-developed plants instead of tall, spindly ones, garden mums must be pinched back at least twice during the early part of the growing season. The first pinch is usually done during the first week in June and the second pinch some 5 to 6 weeks later. Some gardeners make their first pinch on May 30 (Memorial Day) and the second on July 4.

The chrysanthemums grown in the greenhouse, whether for cut flowers or for potted plants, are propagated from stem cuttings, while the outdoor hardy mums may be propagated by either stem cuttings or by plant division. Division can be done late in the fall after flowering has diminished or very early in the spring.

All types of geraniums, genus *Pelargonium*, are easy house plants to grow. They all require an ample diet of direct sunlight, a temperature of 60°F (15°C), and a well-drained soil, which should be kept low in nitrogen and on the dry side. Too much nitrogen, insufficient light, and an over-abundance of water may all or individually contribute to lush growth and few flowers. **GERANIUMS**

Plants grown outdoors and brought inside for the winter must be pinched back to restrict vegetative growth throughout the summer months. If this practice is not followed, the plants will act more like foliage plants than flowering ones.

Propagation of all geranium types is accomplished by stem cuttings. The new hybrids presently on the market are obtained from seed as well as from stem cuttings.

Presently, there are seven different types of geraniums grown in the home and garden. The balance of this section will be devoted to the presentation of each of these types, giving the important facts and characteristics about each type. The potting mix most commonly used for geraniums is two parts loam and one part peat moss, plus some coarse sand for drainage.

Zonals or *Pelargonium zonale* are the common bedding or greenhouse types popular for their ability to flower from early spring to late fall. They are identified by having a horseshoe-shaped zone of dark coloring on most of their varieties. Many varieties also have variegated foliage.

Regals, shows, fancy geraniums, or *Martha Washingtons* have leaves that are palmate, wrinkled, and toothed at the margins. The generic name for this type is *Pelargonium domesticum.* To propagate this type, stem cuttings are taken only during the summer, inserted in sand, and placed in a shaded cold frame until well rooted. Then they are kept in a cool greenhouse throughout the winter, receiving several pinchings to keep the plant compact. During the following spring and summer, these year-old plants will finally flower.

Ivy-leaved geraniums or *Pelargonium peltatum* have slender trailing stems with ivy-shaped leaves. Their growth habit makes them ideal candidates for hanging baskets.

Scented-leaved geraniums have leaves with attractive forms that are strongly scented and aromatic when crushed. This type is more shade-tolerant than the zonals, and, even though their flowers are small and insignificant, they make excellent house plants. The following varieties are the most popular scented geraniums:

P. graveolens	Rose
P. limoneum	Lemon
P. tomentosum	Peppermint
P. fragrans	Nutmeg
P. crispum	Citronella

Dwarf geraniums belong to a species of Pelargonium already mentioned, i.e., zonale. These plants are varieties of zonal geraniums noted for their small size. These plants usually do not exceed 8 to 9 in. (20–22.5 cm) in height.

Miniatures are another variety of zonal geraniums. They are even shorter in height than the dwarf varieties.

Irenes are noted for their large, semi-double flowers, which form in clusters throughout the summer. These plants usually grow to a height of 18 to 24 in. (45–60 cm).

The insect enemies which give a grower of any variety or geranium type the most difficulty are the red spider mites and white flies. These are most easily controlled by using an insecticide on the plants on a weekly basis. Frequent syringing of the foliage also helps to wash the insects off the foliage.

GESNERIADS

The gesneriads are a popular plant family characterized by a wide range of colorful species, producing flowers throughout much of the year. Once it is understood that the growing conditions for all gesneriads are basically the same, i.e., a warm humid atmosphere, bright but indirect light, and a rich, well-drained soil, the homeowner should have very little difficulty in growing and flowering any member of this family.

The most popular gesneriad grown is the African violet, genus *Saintpaulia*. It can be propagated from leaf petiole cuttings, from seed, or by division of large plants. It prefers a low light intensity between 300 and 600 foot-candles and can be grown totally under artificial lights. The plants should be positioned 12–18 in. (30–45 cm) below the fluorescent lights and should receive 12 to 18 hours of light. The formation of yellow leaves indicates too much light, while failure to flower usually indicates insufficient light.

African violets grow well in the temperature range found in most homes, 65–75°F (19–22°C), and will cease thriving when the temperature drops below 60°F (15°C). These plants are best watered from below by inserting a wick in the drain hole of the flower pot or by placing the pot in a saucer of water. Watering from above may result in the occurrence of leaf spot diseases and crown rot.

During periods of active growth, liquid fertilizer should be applied to each plant on a routine schedule of once every two weeks.

Individual African violets must be divided frequently to prevent them from becoming overcrowded; they should be transplanted into a rich, loamy potting mixture having a pH between 6.0 and 6.5.

The balance of this section will consist of a discussion of the many other types of gesneriads grown, giving the methods of propagation and the growing requirements for each type, plus any additional comments that are necessary for their successful growth.

Gloxinias, genus Sinningia, may be grown from seed, tubers, or from leaf cuttings. The seed is extremely fine and will germinate faster when exposed to light. Seed sown in the month of January will produce plants that will flower during the summer months. Gloxinias prefer to receive bright, indirect light, and when they are subjected to low light intensities, their growth becomes both leggy and spindly.

Except for the fact that gloxinias suffer whenever the temperature exceeds 75°F (22°C), all of their other growth requirements are the same as for African violets.

The propagation of gloxinias by leaf cuttings is the same as for rex begonias, mentioned previously in the chapter on propagation. They are usually taken during the spring months, and the resulting plant will flower a year later during the spring months. Gloxinia tubers are planted ½ in. (1.25 cm) below the soil line during the months of November to February and should begin to flower in May. After flowering has ceased, the plant is allowed to dry out and become dormant. It should be stored at 50°F (10°C) until the following January or February, at which time it should be repotted with fresh soil.

Flame violet, genus Episcia, is noted for both its colorful blooms and striking foliage. Its growth requirements are the same as for African violets. New plants are obtained from stem cuttings, leaf cuttings, runners, and from very fine seed. The seed will germinate in 25 to 40 days when exposed to a warm climate between 70 and 80°F (21 and 26°C). Most varieties of Episcia are best displayed and grown in hanging baskets.

The two most common members of the *Columnea genus* are *goldfish plant* and *Norse fireplant*. New plants are obtained from either stem cuttings or from very fine seed. Members of this genus must be grown at a temperature above 55°F (13°C) and require very little water from September to March. They also prefer a soil mixture that is well fortified with organic matter and require fertilization every three weeks during the periods of active growth.

Members of the genus *Achimenes*, commonly called *nut orchid* and *widow's tears*, come in a variety of sizes, shapes, and colors, making them great for hanging baskets. Their culture is very similar to that of African violets. They are propagated from very fine seed, from stem cuttings, and from rhizomes. They bloom profusely during the summer months, but become dormant for four to five months during the fall. During this interval of time, the plants should be stored dry at 60°F (15°C). The rhizomes should also be stored and later repotted to produce plants for the following summer.

The genus *Aeschynanthus*, or *lipstick plant*, is characterized by dark green, waxy leaves, bright red, orange, or yellow tubular flowers, and a trailing growth habit. Varieties of this genus prefer to be grown at high temperatures and high humidity levels.

The genus *Codonanthe* resembles *Columnea* and *Aeschynanthus* in growth habit, but is distinguished from them by the presence of pinhead-sized red spots on the underside of its glossy green leaves. Its flowers are either white or shaded pink.

The genus *Kohleria* is characterized by velvet leaves with interesting patterns and tubular flowers frequently marked with a deeper contrasting color. The flowers on the various species of *Kohleria* may appear in clusters or singly along the stem. The plant may not enter its normal dormant state if cut back directly after flowering has finished. It may be propagated from rhizomes and from stem cuttings.

The genus *Nematanthus*, commonly called *Hypocyrta*, has small flowers that are pouch-shaped on the underside. Some varieties have a trailing growth habit, whereas others are upright. All of its varieties are propagated from stem cuttings.

The two most common species of *Rechsteineria, cardinal plant* and *Brazilian edelweiss*, are both characterized by having tubular flowers that form in clusters at the top of each main stem. Cutting back the stems may prevent dormancy from occurring and ensure continuous flowering. Both species may be produced from either woody tubers or from seed sown in January. Flowers will form on most plants approximately 6 to 12 months after the seeds germinate.

Seeds of *temple bells, genus Smithiantha*, are sown from March through June, producing flowering plants the following winter. This genus is characterized by leaves having a velvet texture and bell-shaped flowers produced on upright stems. New plants are obtained from rhizomes or from seed. These plants grow best at high temperatures and high humidity levels.

Cape primrose, genus Streptocarpus, is propagated by sowing its fine seed, from leaf cuttings, or by crown division. The seeds are usually started in February or March to bloom during the summer months.

The roses grown today in outdoor gardens are one of *six* specific types based on the size of the flowers, the length of their stems, and the natural growth habit of the plant.

Hybrid tea roses are characterized by recurring blooms and a wide range of flower colors and forms, having flowers that range from a single row of petals to the more popular doubles. Most hybrid teas have long pointed buds which yield flowers that are fragrant. Botanically they are a cross between a hybrid perpetual and a tea rose.

Floribundas or hybrid polyanthas are obtained by crossing a polyantha with a hybrid tea rose. Floribundas are versatile in all respects, yielding flowers borne in clusters.

Polyanthas are used for group or border plantings. They form clusters of small flowers that are usually not fragrant. These plants are hardy and vigorous and result from a cross between the dwarf China rose and a Japanese rose.

Grandifloras are obtained by crossing a hybrid tea rose with a floribunda. Their stem lengths fall between that of both parents. Grandifloras are easy to grow, and they bloom freely.

Tea roses originated from the China rose and are known for their continuous blooms and their ability to grow on their own roots. They are not winter hardy in the north, however.

Hybrid perpetuals are vigorous and disease resistant. They are extremely winter hardy and yield fragrant flowers. Many of the flowers, however, may "blue" with age.

Other types of roses grown but not falling into one of the six true classifications are shrub roses, miniatures, tree roses, and climbers.

Shrub roses may be used as hedges, screens, specimens, and in mixed plantings. They are winter hardy and require little maintenance except for an occasional pruning to contain their growth. They also yield attractive fruits in the fall.

Miniatures may be planted in rock gardens or as a border plant and are frequently used for tiny arrangements and corsages because of their very small blooms.

Tree roses are usually composed of three components. The first is an understock chosen for proper root production, and the second is an understock selected because it produces a straight, sturdy cane. This understock, usually the variety De la Grifferaie, is budded onto the first understock, and is referred to as the main stem or standard. The third component of a tree rose is the desired variety, which is budded onto the standard. To give the tree rose a well-balanced flower head, usually two or more buds are used. These plants make a big hit in any garden but do need winter protection.

Climbing roses consist of a group of plants that have upright, long arching canes. Some have large flowers, and others produce small flowers. They do not "climb" but must be trained to climb by being tied to a specific structure.

The planting considerations for all rose plants are:

Planting
Considerations

1. Purchase the best possible plants, because roses should be considered as a long-term investment.
2. Plant the budded or grafted plants in the early spring or in the fall.
3. Position each plant so that its bud union is just below the surface of the soil, at a depth of 1 to 2 in. (2.5–5 cm); see Fig. 37-1.

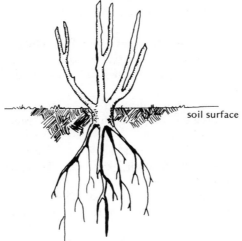

soil surface

FIGURE 37-1
Proper depth to plant a rose.

4. Make certain that the soil is well drained and that it has been well fortified with humus to a depth of 2 ft (0.62 m).
5. Do not allow the roots of the newly selected rose to dry out before planting occurs.
6. Make certain that no air pockets develop in the root area. This is best accomplished by watering and tramping around the plant during the planting operation.
7. Make certain to use the recommended distances between the specific rose types, i.e.,

Hybrid perpetuals	2 to 2½ ft (0.62–0.77 m)
Hybrid teas	1½ to 2 ft (0.46–0.62 m)
Floribundas	1½ to 3 ft (0.46–0.92 m)

Regardless of the rose selected, all roses have specific growing requirements. They need plenty of sunlight, with a minimum of 6 hours per day. They should be planted with a specific purpose or design in mind. They should receive good air circulation to minimize disease infestation, yet be protected from high winds.

Cultural Requirements

The maintenance of established rose plants involves proper watering, fertilization, pruning, and pest control. In certain locations, winter protection is an additional maintenance concern.

Maintenance of Established Plants

Fertilization. Roses need a minimum of two applications of a complete 5-10-5 fertilizer, once during early spring and the second application in midsummer.

Watering. The soil must be kept thoroughly soaked during the entire growing season, but every attempt should be made to keep the foliage dry to minimize fungal infections.

Pest control. A dormant spray of lime-sulfur should be applied in the early spring. Thereafter, the foliage should be covered with a fungicide throughout the entire growing season to minimize mildew and black spot infestations. Sanitation will also help to combat infection.

Aphids and Japanese beetles are the insects causing rose growers the most concern. Spraying the foliage with an insecticide is effective, but care must be exercised to apply only on dry, sunny days. A more desirable method of application is the systemic approach. This involves applying to the soil a water-soluble insecticide, which is absorbed by the rose roots and transmitted throughout the entire plant. Insects, either of the chewing or sucking type, will be destroyed, because the insecticide is ingested into their bodies.

Pruning. This practice is done for two specific reasons: (1) to remove any undesirable wood and (2) to maintain the desired size and shape. On all but the climbers, pruning is usually done in the early spring before the buds begin to swell. Each cane on a hybrid tea should be cut back to within 6–8 in. (15–20 cm) of the ground, three to four buds per cane being left.

Winter protection. No mulch should be applied to any rose until the ground has completely frozen. Soil mounded around the base of the plant is effective against severe winter conditions if applied after the ground surrounding the rose has frozen.

Appendix A

Laboratory Exercises

Laboratory Exercise 1
PINE TREE IDENTIFICATION

PURPOSE

1. To learn how a plant key is used to identify an unknown plant.
2. To become familiar with some of the terminology used in the identification of plants.

MATERIALS

1. Leaves (needles) and cones of several species of the genus Pinus (pine).
2. Ruler.
3. Magnifying glass (optional).

Note. Needles and cones must be paired off by species, assigned numbers, and placed in plastic bags.

PROCEDURE

1. Select one of the unknowns provided by your instructor and record its assigned number in your notebook.
2. With the aid of a ruler, make and record the following observations:
   a. Number of needles in a cluster.
   b. Average length, in inches, of the needles.
   c. Length of the cone, in inches.
   d. Color of the needles, i.e., green or blue-green.
3. The following botanical terms are defined to aid in your identification:
   glabrous—smooth, without hair
   glaucous—light blue-green in color and covered with a white powder, e.g., grapes
   puberulous—covered with fine, short hair or down
   tomentose—closely covered with down or matted hair

4. With the aid of your recorded data, use the attached key to determine the exact species of your unknown.
5. Turn in your results to your instructor.

I. *Gymnosperms*
Leaves needle-shaped, awl-shaped, or scalelike
1. Trees or erect shrubs
  2. Leaves alternate or in clusters (fasicles), needlelike, fruit a cone
    3. Leaves 2–5 in each cluster, usually sheathed at the base   *Pinus*
      4. Leaves in clusters of 5, not sheathed
        5. Branchlets glabrous
          6. Leaves 4–7 inches long, cones 6–10 inches long, branchlets glaucous   *P. griffithi   Himalayan pine*
          6. Leaves 2½–5 inches long, cones 4–8 inches long, branchlets not glaucous   *P. strobus   White pine*
        5. Branchlets puberulous to tomentose
          7. Branchlets thick and tomentose, cones 2½–3½ inches long   *P. cembra   Swiss stone pine*

      4. Leaves in clusters of 3, sheathed at the base
          8. Cones 2–6 inches long, leaves 6–9 inches long   *P. taeda   Loblolly pine*
          8. Cones 1–4 inches long, leaves 3–5 inches long   *P. rigida   Pitch pine*
      4. Leaves in clusters of 2
          9. Leaves 1¼–3 inches long, plant a shrub   *P. mugo   Swiss mountain pine*
        9. Trees
          10. Leaves blue-green, cones 1–2 inches long   *P. sylvestris   Scotch pine*
        10. Leaves green
          11. Buds not resinous, leaves 3–5 inches long, cones 2 inches long   *P. echínata   Shortleaf pine*
          11. Buds resinous, leaves 1½–3 inches long, cones 3 inches long   *P. virginiana   Scrub pine*
          12. Buds brown and resinous
            13. Leaves ½–1½ inches long, cones 1–2 inches long   *P. banksiana   Jack pine*
            13. Leaves 3–6 inches long and dull, cones 2–3½ inches long, leaves thick   *P. nigra   Austrian pine*
              14. Leaves shiny and 4–6 inches long, cones 1½–2 inches long, leaves slender

P. resinosa
Red or Norway pine
12. Buds white, leaves 3–5 inches long
P. thunbergi   *Japanese black pine*

3. Leaves scattered and alternate
2. Leaves opposite, scalelike, awl-shaped or linear   *Juniperus*
1. Low, prostrate creeping shrubs
II. *Angiosperms*
Leaves not needle-shaped
Metric conversion: 1 inch = 2.5 centimeters

Laboratory Exercise 2
HOW TWO POPULAR POTTING MIXTURES ARE PREPARED

PURPOSE
1. To learn the procedures necessary to prepare two potting mixtures suitable for the growth of plants.
2. To learn the basic components of these two mixtures.
3. To become aware that the soil mixture requires pasteurization, whereas all the ingredients in the soil-less mixture are sterile.

MATERIALS
1. Garden soil
2. Sand or perlite
3. Vermiculite
4. Peat moss
5. Metal pails
6. A clean shovel
7. A large clean container or flat surface for mixing
8. An oven
9. A large, shallow metal pan
10. Ground limestone
11. 5-10-5 fertilizer
12. A clean tablespoon

PROCEDURE
Mixture # 1
—containing soil
1. Use the shovel to fill one of the metal pails with garden soil from an area suggested by your instructor.
2. Make certain to remove all large stones and plant material (sod, weeds, and moss) while filling the pail.
3. Fill the second pail with either sand or perlite, and the third pail with peat moss.

4. Pour the contents of the three pails into the large mixing container or onto a clean flat surface.
5. Use the shovel to thoroughly mix the three ingredients.
6. Make certain to thoroughly moisten the entire mixture with water.
7. Place the moistened mixture in the shallow metal pan and position in the oven, which must be preheated to a temperature of 200°F (93°C).
8. Make sure that the soil mixture stays at this temperature for a minimum of 15 minutes.
9. After pasteurization in the oven has been completed, allow the soil to cool down to room temperature. It can then be removed from the metal pan.
10. It is advisable to wait at least one day before using the newly pasteurized soil.

1. Use the clean shovel to fill one of the metal pails with peat moss.   *Mixture #2*
2. Fill a second pail with vermiculite.   *—soil-less*
3. Mix these two ingredients with a clean shovel on a clean surface or in a clean container.
4. Moisten the mixture slightly, and wait until the peat moss has become moist before adding the ground limestone (6 tablespoons per bushel) and the 5-10-5 fertilizer (8 tablespoons per bushel).
5. The mixture can be used as soon as the mixing is completed.

1. What is the reason for adding the sand or perlite to the garden soil mixture?   *QUESTIONS*
2. What is the purpose of the addition of peat moss to both mixtures?
3. Are all soils suitable to support the growth of plants?
4. List the pros and cons for each mixture with respect to cost, ease of preparation, and texture of composition of the particles.

Laboratory Exercise 3
DETERMINATION OF SOIL TEXTURE

1. To learn how to use a simple settling method to determine the texture of a   *PURPOSE*
   specific soil.
2. To become aware that some soil particles, when suspended in water, settle more rapidly than others.
3. To plot the data obtained on the accompanying soil triangle and to be able to predict the exact soil type.

1. One-half cup of sifted soil.   *MATERIALS*
2. 3½ cups of water.
3. A one-quart fruit jar and lid.

4. A metric ruler.

5. An 8-percent Calgon solution (made by dissolving 6 tablespoons of ordinary Calgon powder, sodium hexametaphosphate, in one quart of water).

PROCEDURE

1. Place the soil, five tablespoons of the 8-percent Calgon solution, and the water in the jar.
2. Place the lid on the jar and shake the jar and its contents thoroughly for five minutes.
3. Allow the contents of the jar to settle for 24 hours.
4. Measure the depth of the soil that has settled and record this depth (the *total depth*).
5. Shake the jar again for five minutes and allow 40 seconds for the contents to settle.
6. Measure the depth of soil that has settled and record this depth (the *sand depth*).
7. Allow 30 minutes to elapse before making your third and final measurement of depth. The difference between this depth and that of the sand depth should be recorded as the *silt depth*.
8. The *clay depth* is the amount still in suspension and is determined by subtraction between your total depth and the summation of the sand and silt layers.

An example of the calculations involved is as follows:

Total depth of settled soil	20 mm
Depth of sand layer	8 mm
Depth of silt layer	8 mm
Depth of clay layer (20 − (8 + 8) =	4 mm

$$\% \text{ sand: } ^8/_{20} \times 100\% = 40\%$$

$$\% \text{ silt: } ^8/_{20} \times 100\% = 40\%$$

$$\% \text{ clay: } ^4/_{20} = 100\% = 20\%$$

Reference to the soil triangle shown in Fig. A-1 shows this soil to be a typical loam.

Note. If your soil texture test gives a soil with less than 40 percent clay, consider yourself fortunate, because your soil should be easy to manage and productive.

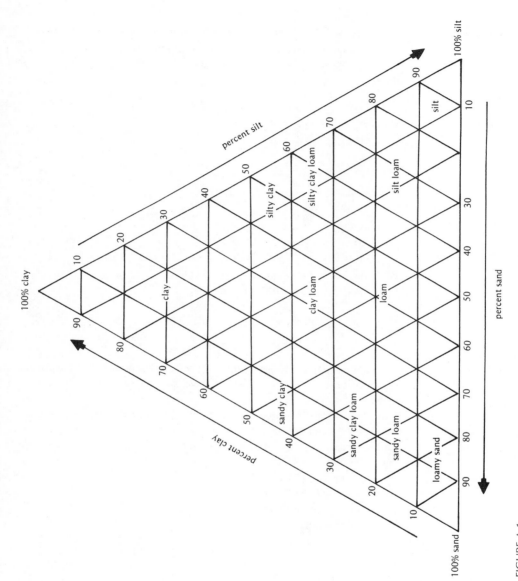

FIGURE A-1
Soil triangle.

215

Laboratory Exercise 4
SOIL TEXTURE

1. To compare the drainage properties of several different soil mixtures.
2. To learn that drainage is related to soil texture or particle size.
3. To introduce the technique of recording observations at specific intervals of time.

MATERIALS 1. Various soil mixtures
   a. Sand
   b. Clay
   c. Soil
   d. Gravel
   e. Peat moss
2. Glass funnels
3. Glass columns—12 in. (30 cm) long
4. Spun glass wool
5. Beakers—250 ml (graduated every 25 ml)
6. Graduated cylinders—100-ml size
7. Water
8. Ring clamps and stands

PROCEDURE 1. For each column to be tested, fill a 100-ml graduated cylinder with water. To conserve time, your instructor will have each column ready to receive the water.
2. Pour the 100 ml of water into the funnel positioned above the column. Be extremely careful not to spill any of the water.
3. On the accompanying data page, record the time that the water was added to each column.
4. At the following prescribed times, record for each column the volume of water in its collection beaker:
   a. 5 minutes after addition
   b. 10 minutes after addition
   c. 15 minutes after addition
   d. 2 hours later
   e. 24 hours later

QUESTIONS 1. Which column permitted the fastest flow of water through it?
2. Which column slowed down the penetration of water through it the most?
3. What is the relationship between particle size and drainage?

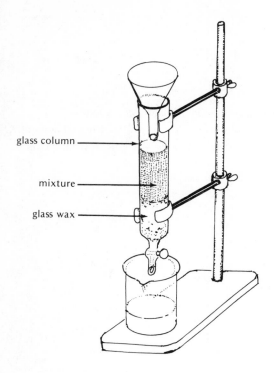

glass column

mixture

glass wax

FIGURE A-2
Soil texture apparatus.

Data Page for Exercise 4
SOIL TEXTURE

| Soil mixture | Starting time | 5 min | Volume of water collected after | | | |
			10 min	15 min	2 hr	24 hr

A COMPARISON OF THE BACTERIAL CONTENT OF
FOUR COMMON GROWTH MEDIA

PURPOSE 1. To learn a simplified technique for the testing of bacterial content.
2. To determine which of the four media would have the lowest bacterial content.
3. To become familiar with the necessity to pasteurize all plant-growing media used indoors.

MATERIALS 1. The four growth media
   a. Compost
   b. Humus from the forest floor
   c. Garden soil
   d. Pasteurized potting soil
2. Glass columns (4), 10 in. (25 cm) long
3. Spun glass wool
4. Distilled water
5. 250-ml beakers (4)
6. Ring stands (4)
7. Clamps (4)
8. Graduated cylinders, 100 ml (4)
9. Eye droppers (5)
10. Prepared petri dishes (5)

PROCEDURE 1. The set-up for this exercise is identical to Laboratory Exercise 4.
2. Place one of the growth media in each of the four glass columns on top of the glass wool.
3. Add 100 ml of distilled water to each column.
4. Allow the water to percolate through each medium.
5. Label each petri dish before it receives a filtrate from a specific column. The fifth dish will serve as a *control* and receive 5 drops of distilled water.
6. Add 5 drops of filtrate from each collection beaker to the correctly labelled petri dish.
7. Place the five dishes in a warm, dark place for 3–4 days.
8. Make the necessary observations at this time to determine which of the four media shows the least amount of bacterial growth and which displays the most.

QUESTIONS 1. Which medium would be best suited for the germination of seeds?
2. Which would be the worst?

HOW TO FERTILIZE A LARGE TREE

1. To learn the procedure commonly used to fertilize large trees.
2. To become familiar with the measurements involved in determining the amount of fertilizer required.

1. 5-10-5 fertilizer
2. Crowbar
3. Tape measure
4. Scales for weighing the fertilizer
5. Pencil and paper for calculations
6. Hammer
7. 20–30 wooden stakes

1. Measure the distance from the tree's trunk to the outer limits of the tree's branches (spread of the tree).
2. Calculate one-third of this distance. The answer is used to determine how far from the trunk the first four stakes should be positioned. Place the stakes at right angles to the tree's trunk.

If the distance from the trunk to the outer limits (radius) is 18 ft (5.5 m), fertilizer is added in the area from this point to within 6 ft (1.84 m) of the trunk (see Fig. A-3).

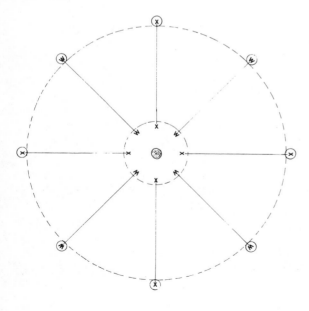

FIGURE A-3
Large tree fertilization.

3. Position the next four stakes at the outer limits of the tree in a direct radial line with the first four stakes positioned.

4. Place an additional four stakes midway between each of the four original stakes (see Step 2) at the correct distance from the trunk.

5. Repeat step 3 by positioning four more stakes at the outer limits of the tree.

Note. At this point in the exercise, there should be a total of 16 stakes positioned, i.e., eight of them at 45-degree angles to each other and at a distance from the trunk of one-third the overall distance, and the remaining eight at the outer limits of the tree on direct radial lines with the inner stakes.

6. Use the crowbar to make holes at intervals of every two feet between each radiating set of stakes. The depth of each hole should be 8 to 10 in. (20–25 cm).

7. After all the holes are made, count them and record the total number in your notebook.

8. Measure the diameter of the trunk, in inches (centimeters), at a height of 4 ft (1.22 m) and record in your notebook.

9. Determine the amount of fertilizer (5-10-5) required for both large deciduous and evergreen trees by multiplying the trunk's diameter, in inches, by two pounds of fertilizer per inch of trunk (or 0.91 k of fertilizer per 2.5 cm of trunk).

10. Determine the amount of fertilizer added to each hole by dividing the weight of fertilizer required by the number of holes.

Laboratory Exercise 7
PHOTOPERIOD

PURPOSE

1. To learn basic potting techniques.
2. To observe the response of poinsettia plants to photoperiod (length of daylight).

MATERIALS

1. Pasteurized potting soil
2. Rooted poinsettia cuttings
3. One 4-inch (10-cm) plastic pot
4. Wood label
5. Black plastic or cloth (for the covering of certain plants)
6. One 24-hour clock timer

PROCEDURE

1. Write your name on a wooden plant label.
2. Select a rooted poinsettia cutting and one 4-in. (10-cm) pot and proceed to the potting area.

3. Place a small quantity of soil in the bottom of the pot.

4. Position the cutting in the center of the pot and add additional soil to secure its position.

5. Press down on the soil in the pot to make sure that no air spaces exist.

6. Place the newly potted plant in a waterproof flat and water it thoroughly.

7. Make weekly visual comparisons between your plant and those receiving light at night.

1. What might happen to the plant if an air space were present?   QUESTIONS

2. Why is it necessary to thoroughly water each newly potted plant?

3. If you owned a greenhouse on a well-lighted highway, how would you care for your poinsettias?

4. Where are the flowers on a poinsettia? Are they large or small?

Laboratory Exercise 8
HYDROPONICS

1. To learn that plants can be grown in a water solution without the presence of soil.   PURPOSE

2. To learn which chemical elements are needed for normal plant growth.

3. To learn the identifying deficiency symptoms when a specific element is lacking.

1. Rooted chrysanthemum cuttings   MATERIALS

2. Cork—size #14

3. Cork borer

4. Jackknife

5. Scotch tape

6. 3 × 5-in. card

7. Gas collection bottle

8. Nutrient stock bottles

1. Using the cork borer, bore a large hole in the center of a size #14 cork.   PROCEDURE
   Then cut the cork into two equal halves with the jackknife.

2. Write your name and specific nutrient solution on a 3 × 5 card and tape it on the outside of the bottle.

3. Fill the gas collection bottle with the specific nutrient solution, leaving some air space in the neck of the bottle.

4. Place the rooted cutting under the water tap and carefully rinse off most of the rooting medium.

5. Position the cutting in the center of the cork. Once the cutting is in position, tape the two halves of the cork together. The taping will allow for ease in removing the cutting while refilling the gas collection bottle.

6. Place the cutting in the gas collection bottle.

7. During the *five* weeks that the exercise will run, replace the solution in each bottle every two days with fresh solution from the correct stock bottle.

8. At the termination of the exercise, observations should be made and comparisons drawn between the different treatments.

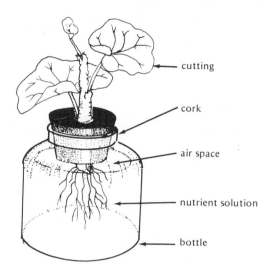

cutting

cork

air space

nutrient solution

bottle

FIGURE A-4
Hydroponics setup.

QUESTIONS

1. What is the importance of the air space in the bottle?
2. Why is it necessary to replenish the solution every two days?
3. What are the elements essential for ideal plant growth?
4. What deficiency symptoms did your plant display?

Laboratory Exercise 9

PURCHASING PLANTS FROM A CATALOGUE

PURPOSE

1. To discover the proper procedure to follow when ordering.
2. To obtain additional knowledge about a specific plant by reading the catalogue.
3. To develop an awareness that any plant purchased will provide both a savings and satisfaction to the purchaser.
4. To learn specific planting instructions with respect to when and how to plant.

1. From the attached list of free catalogues, select the one that interests you the most and send a letter requesting it.

2. In your brief letter, explain that receiving the catalogue is an assignment for your horticulture class.

3. When you receive the catalogue, use the accompanying order form to purchase a specific plant or seed packet. (Limit your purchase to $1.00, unless you desire to spend more.)

4. When you receive your purchase, show it to your instructor for proof of purchase and fulfillment of the exercise.

5. If your selection was a seed packet, containing a large quantity of seeds, they can be planted in a seed flat to be later distributed among your classmates. (With this thought in mind, each member of the class could receive a wide variety of plants to take home.)

Compare the marketing procedures for a given catalogue item, using three different catalogues, with respect to:

1. The unit price.
2. The description of the item, i.e., size, hardiness, growth requirements.
3. What the same item would cost in your immediate area.
4. The minimum handling charge, charge for sales tax, and whether or not the catalogue firm gives a discount for early orders.

Example. Camellia-type tuberous begonias ranged in price from $0.65 to $1.25 for comparable-sized tubers. One catalogue charged $.0.50 for handling the order, while another charged $0.75.

George W. Park Seed Co., Inc.
Greenwood, SC 29647

Seedway Inc.
Hall, NY 14463

W. Atlee Burpee Seed Co.
Warminster, PA 18974

Stokes Seeds Inc.
2046 Stokes Bldg.
Buffalo, NY 14240

Earl May Seed and Nursery Co.
Shenandoah, IA 51603

Kelly Brothers Nurseries Inc.
Dansville, NY 14437

Musser Forests Inc.
Indiana, PA 15701

Stark Brothers Nurseries
Louisiana, MO 63353

Stern's Nurseries
Geneva, NY 14456

J. E. Miller Nurseries
926 West Lake Road
Canandaigua, NY 14424

Spring Hill Nurseries
110 West Elm Street
Tipp City, OH 45371

Laboratory Exercise 10

MEASURING THE PERCENTAGE OF GERMINATION OF SEEDS

PURPOSE

1. To learn a simple procedure to test the germination percentage of seeds.
2. To obtain a rough approximation of the amount of seed to sow to ensure the desired number of plants.

MATERIALS

1. Any available flower or vegetable seeds
2. Petri dishes
3. Blotting paper
4. Marking pen
5. Hand mister

PROCEDURE

1. Cut the blotting paper to fit the bottom of the petri dish. Once it is in position, moisten the paper with the hand mister.
2. Place the seeds to be tested on the moist blotting paper. The size of the seeds will determine the quantity most suitable to add; i.e., use 20 seeds if the seeds are large, and 50 seeds if they are small.
3. Label the cover with the following information:
   a. Seed tested.
   b. The date seeds are sown.
   c. The number of seeds sown.
4. Place the cover on the petri dish and make daily observations, counting the number of germinated seeds.
5. At the end of the week, determine the germination percentage from the following formula:

$$\frac{\text{Number of seeds germinated}}{\text{Number of seeds planted}} \times 100\%$$

Example. If 15 seeds of the original 20 sown germinated, the germination percentage would be 75 percent.

1. To learn the proper techniques for seed germination.
2. To learn how to properly water young seedlings.
3. To become aware of the effect that the use of pasteurized soil has on seed germination.

PURPOSE

1. Seeds—vegetable (tomato) or flower (zinnia)
2. Pasteurized potting soil
3. Unpasteurized potting soil
4. Jiffy seed flats
5. Wooden plant labels
6. Sprinkling can
7. Newspaper

MATERIALS

1. Completely fill one seed flat with the pasteurized potting soil. Fill a second flat with unpasteurized soil.
2. Water each seed flat gently with the sprinkling can until water flows through the drain holes.
3. Once the excess water has left the flats, mark off each seed flat into four long rows with a plant label.
4. Spread the seeds on a large piece of white paper and count out 40 seeds for each flat.
5. Sow ten seeds into each of the four rows in each flat, trying to achieve even distribution of the seeds within each row.
6. Write the date of sowing, the variety planted, and the soil type on a wooden plant label. Place this label at the end of one of the rows.
7. Cover the newly sown seed flats with newspaper. It should be removed only when the flats are watered.
8. You must keep the flats moistened with water while the seeds are undergoing their germination. Exercise care to make certain that the flats are not over-watered.
9. Once germination has occurred and the young seedlings are visible, remove the newspaper covering.
10. Compute the germination percentage for each flat, as presented in Exercise 10.

PROCEDURE

1. What purpose does the newspaper serve in the germination process?
2. Why is it important to completely fill the seed flat with soil?
3. Why is it necessary to label each variety planted and the date sown?
4. Did the use of unpasteurized soil have any marked effect on the germination percentage? If so, what are some of the possible reasons for this difference?

QUESTIONS

Laboratory Exercise 12
EFFECT OF INHIBITORS ON SEED GERMINATION

PURPOSE
1. To learn the effect of various inhibitors on seed germination.
2. To plot on graph paper the comparative results obtained between the experimental control group and the treated group.

MATERIALS
1. Seed packets of bean, cucumber, tomato, and marigold
2. Sterilized sand
3. Graph paper
4. Twenty 4-in. (10-cm) plastic bulb pans (one per treatment)
5. Wooden plant labels
6. 100-ml beakers
7. Sterile glass containers for each inhibitor
8. Inhibitors—tomato juice, apple cider, water extracted from cooked spinach, water extracted from soaked grass seed

Note. Other inhibitors that can be tested are diluted lemon juice, and water extracted from cooked carrots or onions.

PROCEDURE
1. Count out exactly 10 seeds from one of the seed packets.
2. Fill one of the 4-in. bulb pans with sterilized sand.
3. Write your name, the date, the name of seed sown, and the specific treatment the seed will receive on a wooden plant label and place in your pot.
4. Place small-sized seeds just below the sand's surface and the larger-sized seeds at least one inch below the surface.
5. Once the 10 seeds are sown, water your pot with 100 ml of the correct inhibitor.
6. Continue to apply the same inhibitor to your pot for a total of 10 consecutive days, and make daily observations of the number of seeds that have germinated. Record this information in your notebook.
7. If you have been selected for one of the control groups, you are to water your seeds with tap water and record your seeds' germination rate in your notebook.
8. On graph paper, plot the daily germination rate for both the control and inhibitor-treated groups for the same seed. For class uniformity, use the Y axis for the number of seeds germinated and the X axis for the number of days.

QUESTIONS
1. Which of the inhibitors tested suppressed germination the most? The least?
2. Which of the seeds tested seemed to be the least affected by the addition of inhibitors? The most affected?

1. It may be necessary to delegate a responsible student to both treat and record the data for each of the treatments on the weekend. If impossible, all treatments should be left in a cool, low-light environment for the weekend to suppress dehydration as much as possible.
2. To assure uniformity of all inhibitor applications, have each student apply the same volume of liquid, i.e., 100 ml, per day.
3. The exercise as presented suggests the use of four different seeds and five treatments, including the control, to fill 20 pots. If desired, this exercise can be expanded to include other seeds and other inhibitors.

Laboratory Exercise 13

THE EFFECT OF LIGHT ON THE GERMINATION OF SEEDS

PURPOSE

1. To determine which seeds germinate faster in the light than in total darkness.
2. To test the effect that colored filters may have on germination.

MATERIALS

1. Two 40-watt cool white fluorescent bulbs
2. Four petri dishes, numbered 1 through 4
3. Cellophane, one roll red, one roll blue
4. Paper toweling
5. Roll of aluminum foil
6. Seed packets of your choice

PROCEDURE

1. Cut the red cellophane so that it fits snugly into the top of a petri dish. Label this dish number 2. Do the same with the blue cellophane and label the dish number 3.
2. Cut three layers of paper toweling to fit the bottom of each petri dish.
3. Thoroughly saturate the toweling with water and pour off any excess water.
4. Place ten seeds on the toweling in each dish.
5. Using two layers of aluminum foil, cover each dish completely for 24 hours.
6. Dish 1 is to be kept covered throughout the exercise.
7. Dishes 2, 3, and 4 are to receive *one* hour of light daily and then be recovered with the aluminum foil. Dish 4 will be clear with no filter.
8. Once germination occurs in any of the dishes, terminate the exercise and count and record the number of germinated seeds in each dish.

DATA PAGE

Name of seed tested _____
Date exercise started _____
Date exercise was terminated _____

Dish	Number of germinated seeds	% germination
1		
2		
3		
4		

QUESTIONS
1. Was light needed for germination?
2. What effect did the filters have on germination?

Laboratory Exercise 14
THE ROOTING OF LEAF CUTTINGS

PURPOSE
1. To learn the techniques used in the rooting of leaf cuttings.
2. To become familiar with the three main methods used in the propagation of leaf cuttings.

MATERIALS
1. Rex begonia plant
2. African violet plant
3. Propagation flat
4. Sand—rooting medium
5. Hairpins or small pebbles
6. Jackknife
7. Scissors

PROCEDURE
1. Using the jackknife, cut two large leaves from the rex begonia plant.
2. With one leaf, make short cuts through its main veins directly below a point where the veins intersect.
3. Then place this leaf flat on the sand. Use small pebbles or hairpins to hold it in place.
4. Cut the second leaf into V-shaped segments with the scissors. (Each leaf should produce four to six segments.)
5. Insert the point of the V into the sand.
6. Cut several leaves with stems from the African violet plant.
7. Insert the stem of each leaf into the sand.
8. Keep the three types of leaf cuttings both warm and moist at all times.

1. Is it possible to start new plants from the leaves of all plants by the above procedures?
2. Why is it important to have the cuts for the rex begonia leaf cuttings associated with the veins of the leaf?

Laboratory Exercise 15

THE PROPAGATION OF STEM CUTTINGS

1. To learn the proper techniques in the rooting of stem cuttings.
2. To become acquainted with the trial and error method of experimentation as a means of determining which method is best, i.e.,
   a. Is the application of a rotting hormone necessary?
   b. Which rooting medium is the best?

1. Stock plants of geranium or chrysanthemum
2. Rooting media—sand, soil, perlite, vermiculite
3. Rooting hormone
4. Jackknife
5. Propagation flat, divided into four sections
6. Wooden plant labels

1. All the cuttings for this exercise must be cut with a jackknife from the plant tips. Each cutting must be at least 3 in. (7.5 cm) long.
2. Remove the bottom leaves from all the cuttings.
3. This exercise requires 24 cuttings. Half this amount (12) must be dipped in the rooting hormone; the remaining 12 are to be left untreated.
4. The propagation flat to be used must be divided into four equal sections. Each section will contain a different rooting medium.
5. In each of the four sections, place three cuttings that have received the hormone with an additional three cuttings that have not been treated.
6. Make sure that each set of three cuttings is well labeled. (The minimum number of labels is 8.)
7. It is important to keep both the cuttings and the media moist at all times, but not excessively wet.

1. What effect did the hormone application have on rooting?
2. Which rooting medium would you recommend for the rooting of stem cuttings?
3. Why is it necessary to take cuttings for rooting that are at least 3 in. (7.5 cm) long?

Laboratory Exercise 16
PROPAGATION BY TUBERS

PURPOSE

1. To learn how tubers are propagated.
2. To learn the importance of the existence of a bud (eye) in tuber propagation.
3. To grow a food plant to maturity.

MATERIALS

1. Potato
2. Jackknife
3. Potting soil
4. Metal pail [10-gallon (35-l) size]

PROCEDURE

1. Select a potato that has started to sprout (form eyes).
2. Using the jackknife, cut the potato into sections, making certain that each section has at least *two* eyes.
3. Place either small stones or sand in the bottom inch of the pail. Place the potting soil on top of the sand or stones.
4. Plant only *one* potato section in a pail and position it in the soil to a depth of 5 in. (12.5 cm).
5. Once the section has been planted, cover it with at least 2 in. (5 cm) of soil, leaving the top 3 in. (7.5 cm) of the pail empty.
6. During the duration of this exercise, about 3 to 4 months, the plants must receive ample sunlight and must be watered every 3 to 4 days.
7. Once the growth period has ended, carefully remove the contents of the pail.
8. Pay special attention to the presence of new potatoes and their position.

QUESTIONS

1. Why is the sand or stone placed in the bottom of the pail?
2. What might happen if the potato were watered every day?
3. Why was it necessary to bury the potato section instead of placing it on the soil surface?
4. Where were the new potatoes produced?

Laboratory Exercise 17
PROPAGATION BY AIR LAYERING

PURPOSE

1. To learn the techniques and supplies involved in air layering.
2. To learn a suitable way of shortening a leggy house plant.

1. Ficus (rubber plant) or dieffenbachia (dumb cane)
2. Jackknife
3. Sphagnum moss
4. Clear polyethylene plastic
5. Black plastic electrical tape
6. Rooting hormone

1. Make a ball (size of a baseball) out of sphagnum moss that has been previously moistened and place this ball on a piece of absorbent paper toweling.
2. Next, notch the plant stem around its complete circumference, approximately ½ in. (1.25 cm) deep, with the jackknife.
3. Dust the notch with rooting hormone.
4. Use the jackknife to cut the moistened ball of moss into two equal halves.
5. Position the two halves of the moss in the notched area on opposite sides of the plant stem.
6. Cover the moss with the clear polyethylene plastic, and seal both top and bottom with the electrical tape.
7. Treat the parent plant as usual, supplying the necessary water and fertilizer when needed.
8. Once roots are evident, cut the newly rooted plant section away from the bottom portion and plant it in a suitable container.

1. What are some of the advantages to air layering?
2. What are some of the disadvantages?
3. Describe the condition of the sphagnum moss at the termination of this exercise (once rooting has occurred and the plastic has been removed).

Laboratory Exercise 18

PROPAGATION BY GROUND LAYERING

1. To learn the steps to follow to produce new plants by ground layering.
2. To become familiar with the woody plants that easily adapt themselves to ground layering.

1. One of the following plants: forsythia, yew, blackberry, lilac, rhododendron, wisteria, juniper, quince, or hydrangea
2. Jackknife
3. Shovel
4. Peat moss

5. Sand

6. Wood chip

7. Forked twig or staple

PROCEDURE

1. Select any of the plants listed above whose branches will bend to the ground.

2. Using the shovel, loosen the soil in the region where the layering will take place and mix in both sand and peat moss.

3. Cut a notch in the stem before covering it with soil. (The depth of the notch should be about one-third the thickness of the stem.)

4. Place a small wood chip in the notched area.

5. Bury the injured stem section to a depth of 4 in. (10 cm).

6. Position a forked twig or large staple over the stem to ensure that the buried stem remains in place.

7. Finally, make certain that the soil is packed *firmly* in the area that has just been layered.

QUESTIONS

1. Why is it advisable to mix the sand and peat moss with the existing soil?

2. What is the purpose of notching the buried stem?

3. Why is the small wood chip placed in the notched area?

4. Why is it important to pack the soil firmly where the layering took place?

Laboratory Exercise 19

GRAFTING A CACTUS

PURPOSE

1. To learn the procedure to follow in grafting cacti.

2. To become familiar with the terminology associated with grafting.

MATERIALS

1. Rooted cuttings of Pereskia aculeata, a cactus with a tall erect stem (the stock)

2. Another cactus species, whose growth habit can be either round or pendulous (the scion)

3. Corsage pins

PROCEDURE

1. Cut a wedge-shaped piece from the top of the Pereskia cutting (the stock). See Fig. A-5.

2. Cut a thin slice from each of the four sides of the base of the other cactus species selected (the scion).

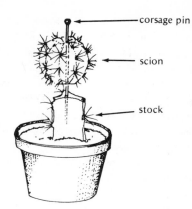

corsage pin

scion

stock

FIGURE A-5
Grafting a cactus.

3. Place the scion on top of the stock and secure in position with a corsage pin.
4. Place the grafted plant(s) in a warm location until the union is complete.
5. The corsage pin can then be removed.

Laboratory Exercise 20

FERN PROPAGATION FROM SPORES

1. To become familiar with one method used in the propagation of ferns.   *PURPOSE*
2. To learn the environmental conditions required for the germination of fern spores.

1. Dried fern fronds containing spores (obtained beforehand from most large   *MATERIALS*
   greenhouse-grown ferns)
2. One 6-in. (15-cm) sterile clay pot
3. One 8-in. (20-cm) sterile plastic pot
4. Peat or sphagnum moss
5. Sterile sand or brick dust
6. Gravel
7. Pasteurized potting soil
8. Plastic bag
9. One 6–7-in. (15-17.5 cm) round glass plate

1. Place a piece of broken pot over the drainage hole of the 6-in. sterile clay pot.   *PROCEDURE*
2. Next, fill the bottom inch (2.5 cm) of the pot with a layer of gravel.
3. Place pasteurized soil on top of the gravel and completely fill the pot.

4. Completely saturate the pot with water. (When the soil is saturated, the excess water will flow through the drain hole.)

5. Press the soil down to eliminate any air spaces and then add a thin layer of sand or brick dust. Moisten this layer.

6. Place the dry fern fronds in a clear plastic bag and shake the bag vigorously to dislodge the spores.

7. Cautiously sprinkle the spores on the surface of the moist sand or brick dust.

8. Place moist peat moss in the bottom of the 8-in. plastic pot before centering the 6-in. clay pot inside it.

9. Fill the entire space between the two pots with additional moist peat moss.

10. Cover the 6-in. pot with the glass plate and place in a warm environment [65–85°F (19–29°C)].

11. Make certain that the peat moss is kept constantly moist throughout the germination period (usually 3–6 weeks).

12. Once germination has occurred, it may be necessary to lift (pick off) some of the sporelings to make room for the remaining ones. They can be planted in either pots or flats containing pasteurized soil.

QUESTIONS

1. What are the similarities and differences between sporelings and seedlings?

2. What is the sequence of steps in the life cycle of a fern?

3. At what stage in its life does a fern produce spores?

HELPFUL TIPS

1. Ripe spores are usually yellow-brown or dark brown in color. The fronds should be placed either in a large paper bag for a few hours or laid flat between layers of newspaper and paper toweling. They can then be transferred to a clear plastic bag.

2. It is advisable to sow spores from several different kinds of ferns to ensure fertilization.

3. The first visible growth is the prothallia stage. Occasional gentle misting will be necessary to have the male element from one prothallium swim to the female part of a different prothallium.

Laboratory Exercise 21
TRANSPLANTING A TREE

PURPOSE

1. To learn the techniques and procedures involved in the transplanting of a tree.

2. To learn the proper method of staking a large tree.

3. (Optional) To plant an evergreen tree that could be decorated at Christmas time.

MATERIALS

1. Tree

2. Shovel(s)

3. Pick
4. Rake(s)
5. Stakes—minimum of three
6. Wire
7. Garden hose—short pieces [6–8 in. (15–20 cm)]
8. Peat moss
9. Turnbuckles—minimum of three (optional)

1. Select the proper site for the plant.
2. Use a pick and a shovel to dig the planting hole. The hole must be both wider and deeper than the ball of earth surrounding the plant.
3. Mix some peat moss with the soil taken from the hole.
4. Carefully place the plant in the hole, making sure that the top of the earth ball is level with the existing soil.
5. Once the plant is level and in position, loosen or completely remove the burlap covering.
6. Shovel the available soil back around the ball of earth and compact it by watering it in or by tramping it in (walking around the circumference of the plant).
7. Use the rake to level off the rest of the available soil.
8. Leave a slight hollow around the outside dimensions of the ball to hold and conserve water.
9. Your instructor may suggest top pruning to diminish the foliage area and thus diminish the rate of evaporation.
10. Guy wires are recommended for trees 10 ft (3.1 m) tall or higher, or for plants placed in a very windy spot.
11. To be successful, at least three guy wires are needed, each extending to a stake 3–5 ft (0.91–1.53 m) away from the plant. Around the trunk of the tree are placed the short pieces of garden hose, through which runs the wire. Turnbuckles may or may not be necessary.

1. What condition does compacting prevent?
2. What might happen to a large tree that was not staked?
3. How long should guy wires be left in place?

Laboratory Exercise 22
TRANSPLANTING SEEDLINGS

1. To learn the proper technique for the transplanting of young seedlings.
2. To learn how young seedlings should be cared for once they are transplanted.

1. Pasteurized potting soil
2. Propagation flat
3. Sprinkling can
4. Wooden plant labels
5. Pencil or dibble
6. Seedlings from Exercise 11

PROCEDURE

1. Seedlings cannot be transplanted (pricked out) and placed individually in a propagation flat unless each one has formed its second pair of true leaves.
2. Fill the propagation flat with the sterilized potting soil.
3. Remove the young seedlings from the seed flat one at a time. This can be effectively done by using a wooden plant label as one might use a shovel.
4. Exercise great care when removing each seedling so that its root system is not damaged.
5. Use a pencil or dibble to make a hole in the soil to accept the seedling.
6. Position the seedlings in the flat so that the space between each one is about two in. (5 cm).
7. Once transplanted, the seedlings must be thoroughly watered but not drenched. Too much water could prove detrimental.
8. After the seedlings have recovered from the shock of being transplanted (2–3 days), they must be checked on a daily basis to determine whether or not they need water.

QUESTIONS

1. What might happen to the newly planted seedlings if their root system were partially damaged?
2. Why is it necessary to wait until each seedling has at least two pairs of leaves before transplanting it?

Laboratory Exercise 23

LAND SURVEY, MEASUREMENT, AND PLOTTING

PURPOSE

1. To become familiar with some of the techniques and equipment employed in land plotting.
2. To learn the importance of using a diagonal measurement to square up any rectangular plot.
3. To learn a simple but effective method of determining the actual slope (grade) of land over a definite distance.

1. Tape measure 50 ft (15 m)
2. Ball of twine
3. Hammer
4. Small hand level
5. Line level
6. Wooden stakes (12 per team)
7. Height stick (6 ft with markings every 3 in. or 2 m with markings every 10 cm)
8. Any available flat surface, i.e., table, photographer's tripod or surveyor's transit

Part A. Measuring and Staking a Perfectly Square Rectangle

1. In each team of four students, two are needed to handle each end of the tape measure, one to record the data, and the fourth to position the wooden stakes.
2. From the following rectangles—10 × 20 ft (3.1 × 6.2 m), 12 × 24 ft (3.7 × 7.4 m), 18 × 36 ft (5.5 × 11 m), 16 × 32 ft (4.9 × 9.8 m)—select the one that you want to measure and inform your instructor before you proceed.
3. Hammer the first stake into the ground. This will be used as the reference point and will be called *stake R*.
4. From stake R, measure the length of the desired rectangle and position at this point a second stake (known as *stake L*).
5. At right angles to both stakes R and L, measure off the desired width, using a minimum of three stakes to form a small arc. Tie the three stakes together with a piece of twine [see Fig. A-6(a)].
6. To measure the other length of the rectangle, use the middle stake of the three involved with the width measurement nearest stake R (call this *stake W*). Another arc is necessary at the terminal end of the length desired. Where the two twines cross, another stake must be placed (call this *stake X*). See Fig. A-7 (b).
7. Stake X may or may not be in the correct position. To determine this, measure the diagonal distance from stake R to stake X and compare it to the distance between stakes L and W.
8. If the distance R-X is greater than L-W, move stake W backwards away from stake X. If the distance R-X is less than L-W, move stake W forward towards stake X. If stake W has to be moved to make both diagonal distances agree, stake X will have to be measured again. See Fig. A-6 (c) and (d).
9. When both diagonals agree, check with your instructor to compare your results to the correct value. See Fig. A-6 (e).

Part B. Determination of a Land Slope

1. Select a site with an apparent gradual slope over a minimum ground distance of 30 ft (9.2 m).
2. At the higher elevation, position the available flat surface. With the use of the hand level, make sure that the flat surface is level.

237

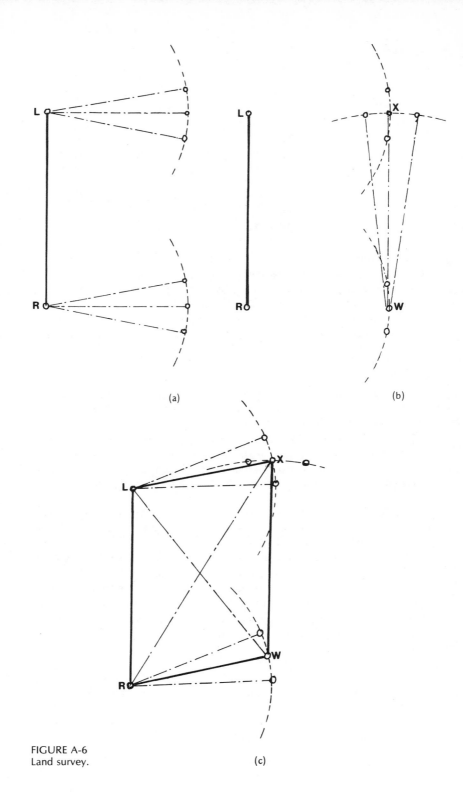

(a)

(b)

FIGURE A-6
Land survey.

(c)

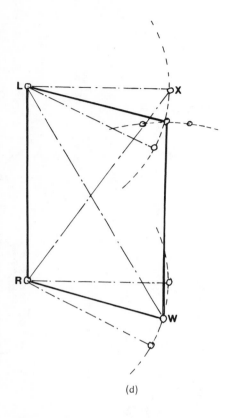

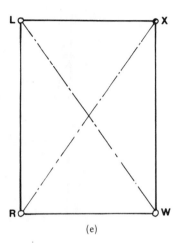

(d)   (e)

FIGURE A-6 (continued)

3. From the flat surface (known as A), measure off a distance of 30 ft (9.2 m). Position the height stick at this point (known as B).

4. Stretch the twine from A to the height stick (B).

5. Place the line level midway between the flat surface and the height stick.

6. Raise or lower the twine on the height stick until the bubble in the line level reaches its desired central position.

7. Record the height where the twine meets the height stick. See Fig. A-7.

8. To calculate the slope of the land, divide the measurement obtained from the height stick by the ground distance. The slope is expressed as: _____ inches per foot or _____ centimeters per meter.

Example

a. If the height stick measures 5 ft, or 60 in. in 30 ft of slope or ground distance, the slope over this distance would be 2 in. per foot.

b. If the height stick measures 160 cm and the ground distance is 8 m, the slope would be 20 cm per meter.

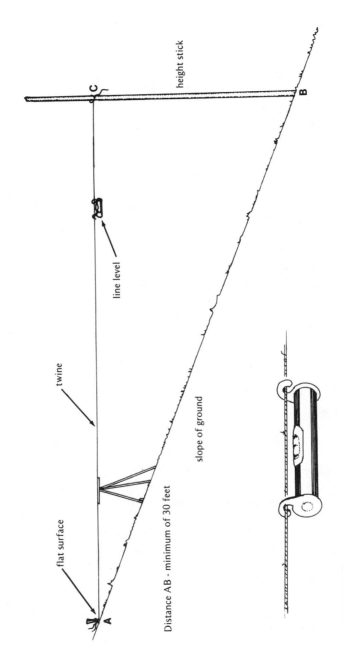

height stick

C

B

line level

twine

slope of ground

flat surface

Distance AB - minimum of 30 feet

A

FIGURE A-7
Land survey.

CREATION OF A LANDSCAPE PLAN

1. To create a scale drawing of a planted area.
2. To realize the importance that measurement, the balance between plants and man-made objects, and knowledge of plant material have to the finished design.

1. Selection of the site area
2. Tape measure, stakes, and hammer
3. Quadrill paper (8 × 11)
4. Number 2 lead pencil
5. Polaroid camera and film (for instructor)

1. From the following list of possible site areas, select the one that interests you the most.
2. The plan that you create can be completely original or can be copied from an existing planting in the nearby area.
3. Your site area must first be staked out and measured with a tape measure before it can be properly scaled down on the final plan.
4. Attempt to make your plan a well-balanced blend between natural and constructional objects.
5. Once you receive this assignment, you will have two deadlines to meet with your instructor. The first (two weeks later) will be to inform him of the following:
   a. The type of site selected
   b. Its specific location, i.e., street number, back yard
   c. Dimensions of the site
   d. Purpose intended for the proposed design

   The second (two months later) will be to turn in your finished scaled plan of the proposed site.
6. The finished copy must be placed on a piece of 8 × 11 quadrill paper, showing the positioning of plant material without actually naming the plants. (If you were a landscape architect, you could supply a plant list for the proposed plan.)
7. The Polaroid camera will be used by your instructor to take a picture of the site you have selected. If it is possible to leave your measurement or boundary stakes in place for the picture, your finished plan will be more meaningful.
8. Figure A-8 is designed to show the simple symbols to be used for the various types of plants, while Figure A-9 is a representative example of a finished plan.

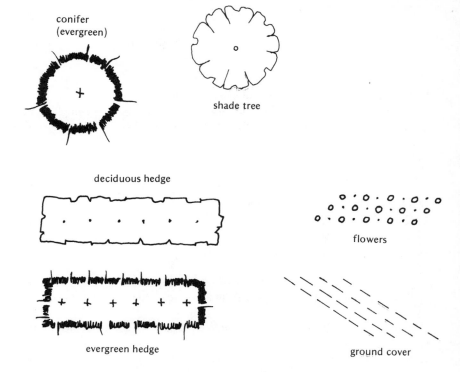

conifer
(evergreen)

shade tree

deciduous hedge

flowers

evergreen hedge

ground cover

FIGURE A-8
Landscape symbols.

Suggestions for possible site areas

1. Foundation planting
2. Herb garden
3. Patio planting
4. Terrace planting
5. Rock garden
6. Flower garden with pathway
7. Roof garden
8. Brook garden (near running stream)
9. Rose garden
10. Play area—putting green, volleyball court
11. Service area—clothes hanging, garbage cans
12. Lawn area—either front or rear
13. Pool planting—either for swimming or display
14. Outside study area near the school

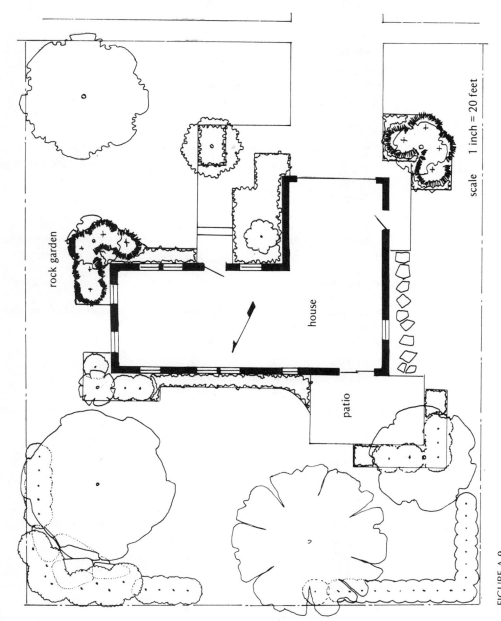

rock garden

house

patio

scale   1 inch = 20 feet

FIGURE A-9
Sample landscape plan.

Laboratory Exercise 25

A COMPARISON STUDY OF SEVERAL LAWN GRASSES
WITH RESPECT TO:

a. Indoor versus outdoor seeding

b. Rate of germination

c. Appearance, i.e., texture, color, thickness

d. Growth rate

PURPOSE

1. To become familiar with the lawn grasses grown in your area.

2. To compare the effects of indoor versus outdoor seeding on the rate of germination.

3. To discover the porosity of clay pots.

MATERIALS

1. Grass seed—a minimum of four types of lawn grasses common to your local area

Example.   For the Northeast: Kentucky bluegrass, fescue, bentgrass, winter rye

2. Four bulb or azalea pans, 8 in. (20 cm) minimum in diameter (either clay or plastic)

3. Four clay pots, 3 in. (7.5 cm) in diameter

4. Pasteurized soil

5. Gravel, for drainage in the large pans

6. Four corks, to plug the drainage holes in the clay pots

7. Wooden labels

8. Marking pen

9. Garden spade or shovel

10. Wooden stakes

11. Four 1-ft-square (0.31-m) areas, each properly labeled and staked off

PROCEDURE

1. Obtain pure grass seed from a local supplier. A small seed packet is sufficient.

2. Place gravel in the bottom of each large pan.

3. Firmly secure a cork in the drainage hole of each 3-in. pot.

4. Place one 3-in. pot in the center of each bulb pan and fill the area between the two pots with pasteurized soil.

5. Carefully sow the soil area in each pot with a specific lawn grass.

6. Use the marking pen to label the pot's exterior and place in the soil area a wooden label containing both the date of planting and the grass sown.

7. Fill the 3-in. clay pot with water and position the pot in a well-lighted area.

8. Spade the outdoor plots to a depth of 8 in. (20 cm) and remove all rocks and visible vegetation.

244

9. Stake off each area and label before sowing the grass seed.

10. Record your findings on the data page.

1. Which grass was the first to germinate?

2. Was there a great difference in the germination time between indoor and outdoor plantings? If yes, give reasons why this occurred.

3. Which grass would you eventually want to sow in your front yard?

LAWN GRASS DATA PAGE

Date sown: _____

Grass	Germination date		Texture (fine or coarse)	Color (light or dark)	Thickness (dense or sparse)	Date grass reaches height of 2 in. (5 cm)
	Indoors	Outdoors				

Laboratory Exercise 26

APPLICATION OF WEEDKILLERS

1. To become familiar with the plot method of obtaining data.

PURPOSE

2. To compare two different methods of application, liquid spray versus solid granules.

3. To learn several broad-leaf weeds, how to count the total number of weeds present in the plot, and how to calculate the percentage killed after application.

1. Tape measure

MATERIALS

2. Stakes—metal or wood

3. Wire or twine

4. Pliers

5. Hammer

6. Liquid 2,4-D weed killer
7. Granular 2,4-D weed killer
8. Liquid applicator

PROCEDURE   **Note.**   The minimum number of plots suggested for this exercise is three. They are

1. The control plot—no application given
2. The liquid 2,4-D plot
3. The granular 2,4-D plot

Additional plots can be used to check the duplication of results if desired.

1. Make the measured plots ten ft (3 m) square. The stakes should be placed in each corner of the plot. Attach the wire or twine to the stakes and then drive them down into the soil with the hammer. Make certain that both the stakes and wire or twine are below the mowing height.
2. To keep the exercise simple, only the two broad-leaf weeds, dandelion and plantain, should be counted in each plot. Other weeds can be included if your instructor wishes to do so.
3. Record the number of each weed in each plot on the data page.
4. After the initial applications have been made, count the weeds in each plot each week for a total of three weeks.
5. Place the liquid 2,4-D weedkiller in the liquid applicator and apply to one of the plots. Repeat the application if rain occurs within 24 hours of the application. This application should be done on a warm day with the temperature at least 70°F (21°C).
6. Do not use the liquid applicator for any other purpose, and when spraying, exercise care not to have the liquid come in contact with any broad-leaf evergreens or flowering plants.
7. Apply the granular 2,4-D weedkiller at the rate listed on the container.
8. To calculate the percentage killed after application, the following formula should be used:

$$\frac{\text{Initial number} - \text{final number}}{\text{Initial number}} \times 100\%$$

QUESTIONS   1. Why should the liquid 2,4-D be applied on a warm day?
2. Why must the liquid application of 2,4-D be repeated if rain occurs within 24 hours of application?
3. Why is this not necessary with the granular 2,4-D?
4. Did you notice a difference in the rate of kill between the two applications of 2, 4-D?

DATA PAGE FOR 2,4-D WEEDKILLER EXERCISE

	Control plot	Liquid 2,4-D	Granular 2,4-D
Before application Number of dandelions			
Number of plantains			
total weeds			
One week later Number of dandelions			
Number of plantains			
total weeds			
Percentage of dandelions killed			
Percentage of plantains killed			
Two weeks later Number of dandelions			
Number of plantains			
total weeds			
Percentage of dandelions killed			
Percentage of plantains killed			
Three weeks later Number of dandelions			
Number of plantains			
total weeds			
Percentage of dandelions killed			
Percentage of plantains killed			

Laboratory Exercise 27
BOW MAKING

PURPOSE
1. To learn the techniques associated with the formation of a bow.
2. To realize that the size or thickness of the ribbon selected is determined by the purpose for the bow, i.e., ½ in. (1.25 cm) for corsages, and 3 in. (7.5 cm) for wreaths, and door and funeral sprays.

MATERIALS
1. Roll of ½-in. (1.25-cm) ribbon
2. Scissors
3. Florist wire, 26 gauge, 8–12 in. (20–30 cm) long

PROCEDURE
1. While you construct your bow, the ribbon may either be left on the roll, or a piece one yard long may be cut from the roll.
2. Start your bow 2–3 in. (5–7.5 cm) from the end and hold the ribbon between your thumb and forefinger [Fig. A-10(a)].
3. Make your first loop of the appropriate size   3–4 in. (7.5–10 cm) long , and make certain to both gather and pinch the ribbon between your thumb and forefinger at the center of the loop.
4. The second loop is formed in the opposite direction from the first loop and must be gathered and pinched in the same manner as the first loop was [Fig. A-10(b)].

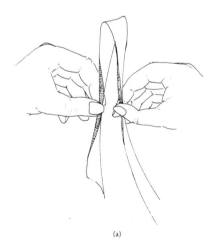
(a)

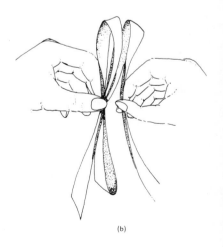
(b)

FIGURE A-10
(a) through (e) Bow making.

248

Note. For those ribbons that have both a shiny and a dull side, it will be necessary to twist the ribbon each time it passes through the thumb and forefinger to keep the shiny surface exposed.

5. Continue making loops until you have three to four loops on each side [Fig. A-10(c)].

6. Use your free hand to place the florist wire over the center of the bow where the loops are held by the thumb and forefinger. Most bow makers push the wire through on the thumb side of the bow [Fig. A-10(d)].

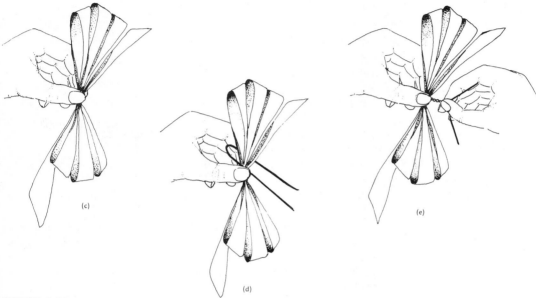

(c)

(d)

(e)

FIGURE A-10 cont.

Note. A length of ribbon can be used instead of the wire to hold the bow together.

7. Push all the loops with your free hand towards the palm of the hand holding the bow. This procedure will enable you to wrap the wire around the center with ease.

8. Make certain that the ends of the wire are even in length before twisting the wire a couple of turns around the center of the bow [Fig. A-10(e)].

9. Once you have properly positioned and tightened the wire or ribbon to the bow, rework the loops to give your bow a more polished appearance.

10. The free ends of the wire or ribbon can then be used to fasten the bow to the corsage.

Note. The procedures listed above for making a corsage bow are identical to those required in the formation of a much larger bow.

SINGLE FLOWER CORSAGE

PURPOSE
1. To learn three basic techniques used by florists: wiring, taping, and bow making.
2. To design a single flower corsage.

MATERIALS
1. Carnation or pompon chrysanthemum flower
2. Asparagus fern
3. Florist wire #18 (medium)
4. Floral tape (roll)
5. Bow material 1/2 to 5/8 in. wide (1.25–1.65 cm)
6. Scissors

PROCEDURE
1. Thread a piece of wire through the calyx of the flower so that both ends of the wire are the same length.
2. Bend the wire down away from the flower head and wrap both ends around the stem.
3. Use the floral tape to cover the wire. Before starting, pull out about 10 in. (25 cm) of the tape from its roll.
4. Start with the tape near the flower head and encircle both the stem and wire with the tape. During this procedure, it is important to keep stretching the tape.
5. Strip off three pieces from the asparagus fern. Each piece should be about 4 in. (10 cm) long.
6. Hold the three fern stems together and arrange them into a fan shape. Next place the taped flower on the fern pieces so that the stems of both the flower and fern touch.
7. Repeat the taping operation, connecting the fern pieces to the flower.
8. Using a minimum of 15 in. (37.5 cm) of bow material, make your bow as previously learned in Exercise 27.
9. Use a small piece of wire to tie the center of the bow material. Once this is done, use the ends of the wire to attach the bow to the taped flower. It may be necessary to use the floral tape to cover this wire.

PURPOSE

1. To learn the techniques necessary for proper floral design.
2. To learn how to arrange the flowers to achieve the design selected (pyramid).

MATERIALS

1. Round plastic container
2. Holding material (oasis, fill-fast)
3. Floral tape (adhesive)
4. Greens: leather leaf
5. Snapdragons: three
6. Pompons: ½ to 1 bunch
7. Jackknife

PROCEDURE

1. Place the holding material in the container. It should be slightly higher than the rim of the container [Fig. A-11(a)]
2. Run two strips of floral tape over the top of the holding material, and attach the ends of the tape to the container.

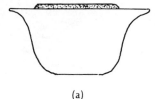

FIGURE A-11
Holiday centerpiece.   (a)

3. Saturate the holding material with water. Pour off any excess water.
4. Separate 3–4 branches of leather leaf into its individual branches. Strip off the bottom two or three leaves.
5. Insert the pieces of leather leaf into the holding material, making certain that they hide the material.
6. The three snapdragons are added next. Cut the stem of each on a slant with a sharp knife. The final length should be about 6 in. (15 cm). Strip the leaves off the bottom 1½–2 in. (3.75–5 cm) of each stem.
7. Position one snapdragon in the center of the holding material in an upright position.
8. Place the other two snapdragons in a horizontal position, one in each end of the holding material [Fig. A-11(b)].
9. Separate the individual flowers from each stem of pompons and arrange them into three piles: (1) fully developed, (2) partially developed, and (3) buds.

10. To achieve the desired pyramidal shape (the base of the design much broader than the top), the flowers near the base must be inserted nearly horizontally, while those near the top will be more vertical [Fig. A-11(c)].

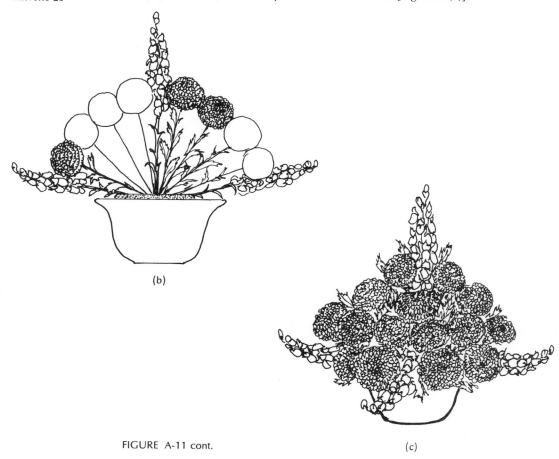

(b)

FIGURE A-11 cont. (c)

11. Make certain to recut each pompon flower on a slant before inserting it in the arrangement. Also, remember to keep turning the arrangement to achieve the proper balance.

12. The first pompons added must define the outer limits of the arrangement's base. Those that follow will be positioned above and inward from the first sequence of blooms and eventually reaching the top of the arrangement. It is important to blend the three types of blooms throughout the arrangement.

13. Add additional branches of leather leaf to cover any exposed flower stems.

DESIGNING A HOLIDAY WREATH

1. To learn how to make an inexpensive holiday wreath. *PURPOSE*
2. To employ some of the basic techniques used by florists.

1. Coat hanger *MATERIALS*
2. Roll of florist wire
3. Evergreen branches—pine, spruce, or fir
4. Hand pruners
5. Pliers
6. Florist wire (medium)
7. Bow material
8. Pine cones, colored glass balls, artificial fruit

1. Bend your coat hanger into a circle. The hook can be left in place to be later *PROCEDURE*
   used to hang the wreath by. If its presence is not desired, it can be removed.
2. Use the hand pruners to cut short sections of evergreen from the large
   branches. A length of 6 in. (15 cm) is suggested for these sections.
3. Arrange these sections into small piles. The number of sections in each pile
   will determine the thickness of the wreath.
4. Tie the end of the roll of wire to the coat hanger.

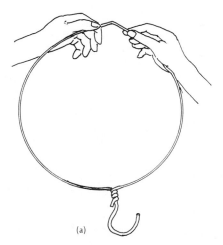

(a)

FIGURE A-12
Holiday wreath.

5. Place a pile of sections on the coat hanger and secure them to the hanger by wrapping the wire around them.

6. Repeat this procedure until you have completely encircled the coat hanger.

7. Use the hand pruners to do the necessary trimming to make the wreath more presentable.

8. To complement the wreath, the following may be added:
   a. A red bow
   b. Pine cones
   c. Colored glass balls
   d. Artificial fruit

(b)

(c)

FIGURE A-12 cont.

9. Wrap florist wire around each of the objects listed above that you plan to use. Use the center of the wire for the wrapping, leaving both ends free.

10. These free ends will be later used to secure the objects to the wreath.

Laboratory Exercise 31

THE COLLECTING, PRESSING, AND DRYING OF GARDEN FLOWERS

PURPOSE

1. To become familiar with the techniques needed to be successful in pressing and drying flowers.

2. To secure the necessary dried plant material to be later used in the exercise involved with designing a floral picture (Exercise 32).

1. Garden flowers and foliage (see the attached list)
2. Jackknife or hand pruners
3. Telephone directories (a minimum of two)
4. Newsprint
5. Paper toweling
6. Facial tissue

1. Gather the plant material on a bright sunny day. Use either the jackknife or hand pruners to remove the desired parts from the plant.
2. Fold a double page of the newsprint in half, so that it fits neatly into the phone book.
3. Place a piece of paper toweling within the folded newsprint so that it lines both the top and bottom folds.
4. Place the facial tissue inside the toweling to ensure a smooth surface.
5. Place the material to be pressed on the right side of the tissue and cover it with the other piece of tissue.
6. Carefully fold the remaining papers (those on the left side) over the plant material.
7. Place the finished *unit* into one of the phone books, leaving ample space between each unit (maximum—10 units per book).
8. It is advisable to place a heavy object on the top of each phone book.
9. To hasten the drying process, the units should be transferred to a dry phone book on a daily basis.
10. It is advisable to keep reliable records for future use, if a large variety of flowers are to be pressed and stored.

Plant Material Adaptable for Pressing

Flowers

Buttercup	Painted daisy
Cosmos	Pansies
Daisies	Pinks
Delphinium	Primula
Dogwood	Queen Anne's lace
English lavender	Scilla
Forget-me-not	Snowdrop
Goldenrod	Sweet alyssum
Gypsophilia (baby's breath)	Sweet clover
Larkspur	Tansy
Marguerite	Wild roses

Foliage

Ajuga	Geranium
Artemisia	Grasses
Bamboo	Herbs
Clover	Juniper
Ferns	Rhododendron

Laboratory Exercise 32
DESIGNING A DRIED FLORAL PICTURE

PURPOSE

1. To learn the techniques necessary to design a floral picture.
2. To allow each student the opportunity to be creative in designing the picture.

MATERIALS

1. Dried plant material—from Exercise 31
2. Small tweezers to handle the dried material
3. Scissors, wire cutters—to trim and shape certain flowers
4. Shadow box picture frame with glass and backing card(s); the most desirable size is 5 x 7 in. (12.5 x 17.5 cm)
5. Background material—velvet, felt, cloth, burlap, etc.
6. Small stiff paint brush
7. Glue—needed if the background is other than velvet to secure the plant material to the background
8. Wire brads

PROCEDURE

1. Cut the background material so that it is about one inch longer and wider than the size of the backing card. At each corner, make a small diagonal cut in the material to avoid bulkiness.
2. Secure the background material to the backing card.
3. Place your desired arrangement initially on a practice space to ensure proper proportion, balance, and design. Figure A-13 shows a representative example of a finished picture.
4. Make certain that the glass for the picture frame is very clean.
5. After you have decided upon your arrangement, carefully pick it up piece by piece with the tweezers and position each piece on the background material. Glue each piece in place.

FIGURE A-13
Dried floral picture.

6. Use the paint brush to make certain that no specks of dust or scattered vegetation remain on the background.

7. Place the picture frame over the arrangement and, exercising extreme care, secure the backing card(s) to the frame with the small wire brads.

8. It is advisable to tape the underside of the frame to prevent dust from entering the finished picture.

POTTING AND FORCING PAPER WHITE NARCISSUS

1. To learn what conditions are necessary to flower paper white Narcissus bulbs.   *PURPOSE*

2. To show how easily the flowering of some plants can be accomplished when certain simple rules are followed.

1. Paper white Narcissus bulbs   *MATERIALS*
2. Washed gravel
3. Wooden plant labels
4. 3-in. (7.5-cm) plastic pots

1. Write your name on a wooden plant label.   *PROCEDURE*
2. Select one bulb and one 3-in. plastic pot and proceed to the area where the washed gravel is.
3. Place a small quantity of gravel in the bottom of the pot.
4. Position the bulb in the center of the pot and place additional gravel around it. Allow the tip of the bulb to project slightly above the pot.
5. After potting, place each pot in a watertight flat. Make certain that the water level in the flat is near the base of the bulb at all times.
6. Avoid placing the bulbs in direct sunlight.
7. Flowering should occur within four weeks of potting.

1. Each bulb was placed in the dark for a period of three weeks before you   *QUESTIONS*
planted it. Why was this necessary?
2. What do you suppose might happen if the water level were not maintained in the flat?

Laboratory Exercise 34
STORAGE AND FORCING OF TULIP BULBS

PURPOSE
1. To learn the various stages that flower buds go through before flowering occurs.
2. To learn the requirements (time and temperature) for forcing certain bulbous plants (man's way of speeding up flower development).
3. To have you recognize the commercial importance of flowering plants out of season.

MATERIALS
1. Pasteurized potting soil
2. 3-in. (7.5-cm) plastic pots
3. Tulip bulbs
4. Wooden plant labels
5. Refrigerator
6. Two plastic bags
7. One 3 x 5 card

PROCEDURE
1. Place 21 tulip bulbs in a plastic bag. It is important to add to the bag a 3 x 5 card containing the date and the day of the week. Place this bag in the refrigerator.
2. Place seven more tulip bulbs in a second plastic bag. Keep this bag in a safe place at room temperature.
3. After six full weeks of storage in the refrigerator, remove three bulbs from the plastic bag and pot these three in 3-in. plastic pots.
4. At the same time, remove *one* bulb from the bag left at room temperature and pot it in a 3-in. pot.
5. Make certain to label each pot and its treatment at the time it is planted.
6. Thoroughly water each newly planted bulb before placing it in a darkened area for the necessary root development. Once roots are apparent through the drainage hole, the bulbs can be forced in full sunlight.
7. Repeat the above procedure each ensuing week until all the bulbs have been planted. (Minimum storage is 6 weeks; maximum is 12 weeks.)
8. Make the necessary observations and record your findings in your notebook. This must be done on a weekly basis.

QUESTIONS
1. What effect did the period of cold storage have on the flowering of the tulip bulbs?
2. In your observations, did you notice any correlation between the length of cold storage and the time involved before flowering occurred?
3. What happened to those bulbs that did not receive a cold treatment?

	STORAGE TIME (NUMBER OF WEEKS)						
	6	7	8	9	10	11	12
Date of planting							
Appearance of bulb at time of planting							
Appearance of plant after planting							
1st week							
2nd week							
3rd week							
4th week							
5th week							
6th week							
7th week							
8th week							
9th week							

Laboratory Exercise 35
A COST ACCOUNTING SURVEY

PURPOSE

1. To become aware of all the factors that must be accounted for when cost accounting methods are used.
2. To realize the importance that cost accounting has in arriving at the final sales price.
3. To discover the margin of profit that can be expected.

BACKGROUND INFORMATION

In our present times of ever-increasing costs, most horticultural operations have turned to the system of cost accounting as a satisfactory method used to determine what they must eventually charge for their products.

If used properly, this method can show the owner what the actual cost was for producing each specific item that he sells. To obtain this figure, it is first necessary to account for the monies spent in each of the following areas:

1. Labor—the amount paid in wages for actual work done.
2. Materials—expenses involved with all those items directly used in connection with the saleable item.
3. Overhead—all additional costs not listed under labor or materials.

PROCEDURE

1. This laboratory exercise requires you to personally contact a specific horticultural business involved in the sale of a product.
2. It is important for you to make an appointment with a specific individual, either by phone or in person, explaining what your intended purpose is. This will allow the person time to gather the information that you seek in order to compute the margin of profit for a specific item.
3. The accompanying work sheet is designed to make you aware of all the factors that must be considered before a final price is arrived at, and is flexible so that it can be used for a variety of differing operations.
4. The work sheet should be used *only* as a guide to aid you in the preparation of the questions you intend to ask. *Do not* spoil your interview by simply requesting that the individual being interviewed complete the sheet.
5. It must be understood that sales from many horticultural operations are seasonal, covering only a small portion of the year. However, the time necessary to prepare for the harvest may require working during the rest of the year. This must be noted when the actual profit for a given product is computed.
6. Because of the difficulty in calculating the effect that land taxes, bank loans, and capital investment have on the total cost picture, these three items should not be considered when the margin of profit is presented.

Saleable product: _____

Time required for product: _____ (days, weeks, months, years)

Total number of saleable units involved: _____

Number of employees needed for production: _____

Land area involved: _____

Cost of:	JanFeb	MarAp	MayJu	JulAug	SepOct	NovDec	total
A. Labor							
B. Materials							
a. plants							
b. containers							
c. machinery							
d. fertilizers							
e. pesticides							
f. tools							
g. related supplies							
C. Overhead							
a. heat							
b. electric							
c. phone							
d. water							
e. insurance							
f. delivery							
g. maintenance							
h. cleanliness							

Once your interview is completed, hand in to your instructor the following information:

Student's name _____

Establishment visited _____

Saleable item discussed _____

Selling price _____

Total monies spent _____

Margin of profit _____

Laboratory Exercise 36
EFFECT OF GROWTH REGULATORS ON PLANT GROWTH

PURPOSE

1. To compare the effect of a single application of a growth stimulant to that of a growth inhibitor.
2. To compare the results of these two applications to the untreated plants (the control group).
3. To make comparisons between leaf growth, and plant height, and to record this data in tabular form.
4. To learn the importance of measurement as opposed to simple casual observations.

MATERIALS

1. Nine plants—geranium, poinsettia, or chrysanthemum
2. Gibberellic acid solution (10^{-4} molar or 35 mg per liter).
3. B-Nine solution (6 oz per gallon of water or 47 ml per liter)
4. Jackknife
5. Chemical balance
6. Plant labels
7. Centimeter ruler
8. Sprayer for the GA solution
9. One 250 ml. graduated beaker

PROCEDURE

1. Separate the nine plants into three equal groups. One group of three will receive an application of gibberellic acid (GA), a second group will be treated with B-Nine, and the third group will serve as the control group.
2. Write today's date on each of the nine labels. On three of the labels, write the notations GA-1, GA-2, and GA-3 and place these three labels on the plants that will be sprayed with GA.
3. Write the notations B-Nine-1, B-Nine-2, and B-Nine-3 on three more labels and place one of them in each of the three pots that will receive a soil drench of B-Nine.
4. On the three remaining labels, write the notations Control-1, Control-2, and Control-3.
5. Use the centimeter ruler to make the initial measurements, and record these on the data page. Each of the nine plants must be measured in terms of (a) the total number of leaves, and (b) the height at the time of application.
6. Spray the gibberellic acid uniformly over each of the three plants labeled with the notation GA.
7. Apply 250 ml. of the B-Nine solution as a soil drench to each of the three B-Nine plants.
8. Allow from 6 to 8 weeks after the application before recording the final conditions.
9. Record the final conditions on the data page, using the same procedures employed in obtaining the initial measurements.

10. For simplicity, compute a simple mathematical average for each test for each group. Then by subtraction, the differences between the initial and final conditions can be found.

11. Make the necessary comparisons between the three treatments.

DATA PAGE FOR LABORATORY EXERCISE 36

Test A. *Total number of leaves* (initial number should be 10 or less)

	Control group		GA group		B-Nine group	
	Initial	Final	Initial	Final	Initial	Final
1.						
2.						
3.						
Average						
Difference						

Test B. *Plant height* (from soil level to the top of the plant)

	Control group		GA group		B-Nine group	
	Initial	Final	Initial	Final	Initial	Final
1.						
2.						
3.						
Average						
Difference						

QUESTIONS

1. When compared to the control plants, what differences were displayed by the plants sprayed with gibberellic acid?

2. What effect did an application of B-Nine have on the plants?

3. What results would you as a commercial plant grower attempt to achieve by using either the gibberellic acid or B-Nine?

Laboratory Exercise 37
DESIGNING A TERRARIUM

PURPOSE
1. To learn how to design a popular gift item that contains living plants.
2. To become familiar with the type and size of plants used in a terrarium.

MATERIALS
1. Pea-sized gravel
2. Granular charcoal
3. Pasteurized potting soil
4. Glass container with lid
5. Suitable plant material—supplied by your instructor

PROCEDURE
1. Select any glass or plastic container that has a top or cover.
2. Place the pea-sized gravel in the bottom of the container (for drainage).
3. Add the charcoal next, placing it on top of the gravel. (The purpose of the charcoal is to keep the container sweet.)
4. Place the potting soil on top of the charcoal. The depth of the container will determine the quantity of soil needed.
5. Next, thoroughly moisten the soil, but do not make it muddy.
6. Before the plants are planted in the soil, it is advisable to clean the inside surface of the container.
7. In your design, make certain that the taller plants are placed in the rear of container and the smaller ones in the foreground.
8. Once the planting has been completed, the terrarium should be placed in a well-lighted setting, but not in direct sunlight.
9. When the top cover is put in position, both the interior temperature and the moisture content will become balanced.
10. If the container clouds up with moisture, this is easily corrected by opening the cover for a short period of time.

QUESTIONS
1. Why is it necessary to plant only small, slow-growing plants in your terrarium?
2. Is the humidity within the terrarium high or low?

APPLE HEAD DOLLS

1. To carve an apple into a human head (an anatomical study). *PURPOSE*
2. To learn how to treat the apple to preserve it.

1. One large apple (Golden or red Delicious are the best.) *MATERIALS*
2. A stainless steel knife
3. Salt in a shaker
4. Lemon juice
5. Small artist's brush
6. Pipe stem cleaner

1. Peel the apple in a continuous strip if possible. Leave the stem intact. Select *PROCEDURE*
   the best side of the apple for the face.
2. Divide the face into three equal parts. Mark each division with the knife.
   [Fig. A-14(a)].
3. At the completion of *each* of the following numbered steps, sprinkle salt on
   the head and brush in the lemon juice. All sharp edges must be rounded with
   the knife.
4. The first anatomical areas to be defined are the *temple-cheek* area
   and the *lower jaw*. Each of the areas involves both a vertical and a
   slanted cut [Fig. A-14(b)].
5. To form the *forehead,* make a vertical cut from the top to the first
   division [Fig. A-14(c)].

front view
(a)

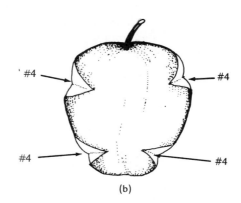

(b)

FIGURE A-14
(a) through (d) Apple head dolls.

265

6. To form the *nose*, three separate cuts must be made. The first is a diagonal cut started ¼ in. above the lower division and sloping upward to meet the forehead. The second cut is a shallow horizontal one into the lower dividing line. This defines the *tip of the nose*. The third cut is a vertical one upward to meet the second cut. This defines the *mouth area* [Figure A-14(c)].

7. To define the *chin area* for a young face, make a horizontal cut into the lower jaw area and complete by making a vertical cut upwards [Fig. A-14(c)]. For an older face, make one diagonal cut in the same area.

8. Make a diagonal cut from the lower division to the base of the apple to define the *back of the head* [Fig. A-14(c)]

9. Make a shallow cut on each side of the bridge down to the tip of the nose to define each *side* of the nose. Make two additional shallow cuts from the tip of the nose to the chin. These define the *sides of the mouth* [Fig. A-14(d)].

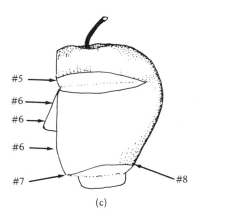

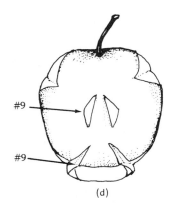

FIGURE A-14 cont.   (c)   (d)

10. For a head with exposed *ears,* start by forming a rectangular area in the middle region of the profile [Fig. A-14(e)]. Exercising great care, continue to shape each ear into its characteristic half-moon shape.

11. To form the *mouth,* make a narrow V-cut in the area below the nose [Fig. A-14(e)].

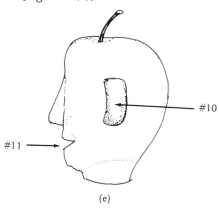

FIGURE A-14 cont.   (e)

12. Remove the stem and the top part of the apple core, leaving the blossom end intact. Punch a hole from the open stem end through the blossom end.

13. Push a pipe cleaner through the hole and twist it at the blossom end. Form a hanger in the top end so that the finished apple head can be hung to dry.

DECORATING A PINE CONE

1. To prepare a simple holiday ornament.

2. To allow each student the possibility of creating his own design.

3. To blend several floral items into one finished product.

PURPOSE

1. White glue

2. Pine cone(s)

3. Straw flowers (minute in size; in a variety of colors)

4. Raffia

5. Scissors and/or a jackknife

6. Sandpaper

MATERIALS

1. Place glue on one end of the raffia and insert it into the base of the pine cone. Bend the raffia into a circular shape so that it gives a "halo" appearance above the pine cone.

2. Cut off any excess raffia before gluing and inserting the other end into the pine cone.

3. Next, count the number of open scales on your cone. (This number will be the same as the total number of straw flowers that you will need.)

4. Select from the variety of colors available the number of straw flowers that you need. Remember, the choice is entirely yours, and you need one for each open scale.

5. Cut off the stems of each straw flower with a jackknife or scissors.

6. Glue a straw flower head on the tip of each open scale. (This is best accomplished by placing a small quantity of glue on the scale.)

7. Since it is not known whether the pine cone will be hung by the raffia "halo" or placed on a flat surface for display, it is necessary to use sandpaper to sand the base of each cone so that it will not tilt to one side when placed on a flat surface.

PROCEDURE

Laboratory Exercise 40

REFORESTATION

PURPOSE
1. To learn the techniques used in planting evergreen seedlings.
2. To become acquainted with the reasons associated with reforestation, i.e.,
   a. To reduce the possibility of soil erosion.
   b. To beautify the area.
   c. To provide a possible home for wildlife.
   d. To provide a source of revenue, e.g., lumber, Christmas trees, or pulp.

MATERIALS
1. Mattock(s) or pick(s)
2. 50-ft (15.3-m) tape measure(s)
3. Wooden stakes
4. Hammer(s)
5. Evergreen seedlings: Douglas fir, pine, spruce

PROCEDURE   **Note.**   The planting of evergreen seedlings for reforestation purposes is designed to be simple and quick. A planting hole is made with a pick or mattock, and a seedling is placed in this hole. It is not necessary that the seedling be planted erect. Once it is placed in the hole, the planting procedure is completed by simply stepping down with your shoe on top of the hole. This is important to prevent damage due to air pockets.

1. The actual planting site will be selected by your instructor.
2. The first student must stretch out the tape measure completely (50 ft or 15.3 m).
3. The second member of the team is to place a wooden stake every 6 ft (2 m) along the tape. (The stakes can be laid flat on the soil or driven into the soil with a hammer.) Once the stakes have been placed, the tape measure can be picked up and a new planting line can be staked out.
4. The third student's job is to use either the pick or the mattock to open a hole in the soil. [The hole should be 3–4 in. (7.5–10 cm) deep.]
5. Because most of the seedlings are small and tend to stick together, a fourth member of the team is needed to separate the seedlings and drop one near each planting hole.
6. The fifth student does the actual planting. (His only concern is to make certain that the roots are placed in the hole.)

268

Careers in Horticulture and Related Fields

Career	Description	Degree	Additional information
Agronomist	Concerned with the interaction between soils and plants	B.A. or Advanced	American Institute of Biological Science 3900 Wisconsin Ave. N.W. Washington, D.C. 20016
Analytical chemist	Analyzes soils, plants, etc. for contaminants	B.S.	American Chemical Society 1155 Sixteenth St. N.W. Washington, D.C. 20036
Arborist	Deals with tree care and maintenance	B.A.	National Arborists Assn. 2011 Eye Street N.W. Washington, D.C. 20006
County extension agent	Helps growers with their plant problems	B.A.	Federal Extension Service U.S. Dept. of Agriculture Washington, D.C. 20250
Ecologist	Relates people, organisms to their environment	B.A. or Advanced	American Inst. Biol. Science (see Agronomist)
Entomologist	Concerned with insects and their control	B.S. or Advanced	Amer. Inst. Biol. Science (see Agronomist)
Floral designer	Deals with flower arranging and retail florist shop	Some college training is beneficial	Society of American Florists 901 North Washington St. Alexandria, VA. 22314
Forester	Deals with all aspects of timber and wildlife management	B.A.	American Forest Institute 1619 Massachusetts Ave N.W. Washington, D.C. 20036
Forestry aide	Helps the forester	Some college	U.S. Dept. of Agriculture Forest Service Washington, D.C. 20250
Grower (floriculturalist)	Production of floral crops	2 or 4 years college	American Society of Horticultural Science P.O. Box 109 St. Joseph, MI. 49085
Landscape architect	Designs home and business landscapes	B.A. or Advanced	American Society of Landscape Architecture 1750 Old Meadow Road McLean, VA. 22101

Career	Description	Degree	Additional information
Microbiologist	Studies bacteria, viruses, and fungi	B.S. or Advanced	Amer. Inst. of Bio. Sci. (see Agronomist)
Nurseryman	Grows and propagates trees and shrubs	2 to 4 years college	National Landscape Nurserymen's Assoc. 832 Southern Bldg. Washington, D.C. 20006 American Assoc. of Nurserymen 835 Southern Building Washington, D.C. 20005
Ornamental horticulturalist	Involved with the growth and culture of all cultivated plants	2 to 4 years college	Amer. Soc. for Horticultural Science 615 Elm Street St. Joseph, Ml. 49085
Pathologist	Studies plant diseases, causes and control	Advanced	Ecological Soc. of America Dept. of Botany University of N. Carolina Chapel Hill, N.C. 27514
Pomologist	Involved with the growth of all types of fruit	B.A. or Advanced	Amer. Soc. of Hort. Sci. (see Grower)
Range manager	Properly manages existing range land	B.A.	Bureau of Land Management Denver Service Center Federal Bldg. 50 Denver, CO. 80255
Sanitarian	Environmental health	B.A.	Amer. Public Health Assoc. 1015 18th Street N.W. Washington, D.C. 20036
Soil conservationist	Involved with all phases of soil conservation	B.A.	U.S. Dept. of Agriculture Washington, D.C. 20250
Soil scientist	Studies all aspects of soils	B.A.	U.S. Dept. of Agriculture Washington, D.C. 20250
Turfgrass production	Concerned with sod production and grass maintenance	B.A.	Amer. Soc. of Hort. Sci. P.O. Box 109 St. Joseph, Ml. 49085
Urban planner	Involved with the renewal of cities	B.A. or Advanced	American Institute of Planners 917 15th Street N.W. Washington, D.C. 20005

State Agricultural Colleges and Extension Services

State residents should direct their questions about plants and related topics to the following addresses. Each of the addresses listed is that of a state agriculture college, most of which are combined with an extension service for the answering of questions, and an experimental station for research purposes. Your letters should be entitled "Agricultural Information" in care of the following:

State	College or University	Address
Alabama	Auburn University	Auburn, Ala. 36830
Alaska	University of Alaska	College, Alaska 99701
Exp. Sta.		Palmer, Alaska 99645
Arizona	University of Arizona	Tucson, Ariz. 85721
Arkansas	University of Arkansas	Little Rock, Ark. 72203
Exp. Sta.		Fayetteville, Ark. 72701
California	University of California	Berkeley, Calif. 94720
Exp. Sta		Davis, Calif. 95616
		Los Angeles, Ca. 90024
		Riverside, Ca. 92502
Colorado	Colorado State University	Fort Collins, Colo. 80521
Connecticut	University of Connecticut	Storrs, Conn. 06268
Exp. Sta.		New Haven, Conn. 06504
Delaware	University of Delaware	Newark, Del. 19711
Florida	University of Florida	Gainesville, Fla. 32601
Georgia	University of Georgia	Athens, Ga. 30602
Hawaii	University of Hawaii	Honolulu, Hawaii 96822
Idaho	University of Idaho	Moscow, Idaho 83843
Illinois	University of Illinois	Urbana, Ill. 61801
Indiana	Purdue University	Lafayette, Ind. 49707
Iowa	Iowa State University	Ames, Iowa 50010
Kansas	Kansas State University	Manhattan, Kans. 66502
Kentucky	University of Kentucky	Lexington, Ky. 40506
Louisiana	Louisiana State University	Baton Rouge, La. 70803
Maine	University of Maine	Orono, Me. 04473

State	College or University	Address
Maryland	University of Maryland	College Park, Md. 20742
Massachusetts	University of Massachusetts	Amherst, Mass. 01002
Michigan	Michigan State University	East Lansing, Mich. 48823
Minnesota	University of Minnesota	St. Paul, Minn. 55101
Mississippi	Mississippi State Univ.	State College, Miss. 39762
Missouri	University of Missouri	Columbia, Mo. 65201
Montana	Montana State University	Bozeman, Mont. 59715
Nebraska	University of Nebraska	Lincoln, Nebr. 68503
Nevada	University of Nevada	Reno, Nev. 89507
New Hampshire	University of New Hampshire	Durham, N.H. 03824
New Jersey	Rutgers University	New Brunswick, N.J. 08903
New Mexico	New Mexico State University	Las Cruces, N. Mex. 88001
New York	Cornell University	Ithaca, N.Y. 14850
Exp. Sta.		Geneva, N.Y. 14456
North Carolina	N. Carolina State Univ.	Raleigh, N.C. 27607
North Dakota	N. Dakota State University	Fargo, N.Dak. 58102
Ohio	Ohio State University	Columbus, Ohio 43210
Oklahoma	Oklahoma State University	Stillwater, Okla. 74074
Oregon	Oregon State University	Corvallis, Oreg. 97331
Pennsylvania	Penn. State University	University Park, Pa. 16802
Puerto Rico	University of Puerto Rico	Rio Piedras, P.R. 00928
Rhode Island	University of Rhode Island	Kingston, R.I. 02881
South Carolina	Clemson University	Clemson, S.C. 29631
South Dakota	South Dakota State Univ.	Brookings, S. Dak. 57006
Tennessee	University of Tennessee	Knoxville, Tenn. 37901
Texas	Texas A. & M. University	College Station, Tex. 77843
Utah	Utah State University	Logan, Utah 84321
Vermont	University of Vermont	Burlington, Vt. 05401
Virginia	Virginia Polytechnic Inst.	Blacksburg, Va. 24061
Washington	Washington State University	Pullman, Wash. 99163
West Virginia	West Virginia University	Morgantown, W.V. 26506
Wisconsin	University of Wisconsin	Madison, Wisc. 53813
Wyoming	University of Wyoming	Laramie, Wyo. 82070

Plant Societies

Some plant enthusiasts' interest in plant material may eventually focus upon one specific plant, a plant family, or one specific plant type. They then desire more detailed knowledge about their particular plant area and frequently join one of the following plant societies to share with others their common interests. Since most correspondence is performed by the secretary of the organization, your initial letter should be addressed accordingly. The initial information that you receive should include requirements for membership, yearly dues, publications available, and interesting research presently being done or contemplated.

Society Name	Address
African Violet Society of America	P.O. Box 1326, Knoxville, Tenn. 37901
American Begonia Society	10692 Bolsa St. Apt. 14, Garden Grove, Calif. 92643
American Bonsai Society	228 Rosemont Avenue, Erie, Pa. 16505
American Boxwood Society	P.O. Box 85, Boyce, Virginia 22620
American Camellia Society	P.O. Box 1217, Fort Valley, Georgia 31030
American Daffodil Society	89 Chichester Road, New Canaan, Conn. 06840
American Dahlia Society	1649 Beech Avenue, Melrose Park, Pa. 19126
American Fern Society	c/o James Caponetti, University of Tennessee, Knoxville, Tenn. 37916
American Fuchsia Society	Hall of Flowers, Golden Gate Park, San Francisco, Calif. 94122
American Gloxinia and Gesneriad Soc.	c/o Mrs. J. Rose, P.O. Box 174, New Milford, Conn. 06776
The American Gourd Society	Box 274, Mount Gilead, Ohio 43338
American Hemerocallis (Daylilies)	c/o Mrs. Arthur Parry, Signal Mountain, Tenn. 37377
American Iris Society	6518 Beachy Avenue, Wichita, Kansas 67206
American Ivy Society	Nat'l Center for Amer. Horticulture, Mt. Vernon, Virginia 22121

Society Name	Address
American Orchid Society	Botanical Museum of Harvard Univ., Cambridge, Mass. 02138
American Peony Society	250 Interlachen Road, Hopkins, Minn. 55343
American Plant Life Society	c/o Dr. Thomas W. Whitaker, P.O. Box 150, La Jolla, Calif. 92037
American Primrose Society	c/o Mary Speers, 202 Champion Street, Steilacoom, Wash. 98388
American Rhododendron Soc.	c/o Mr. R. Berry, 617 Fairway Drive, Aberdeen, Wash. 98520
American Rock Garden Soc.	3 Salisbury Lane, Malvern, Penn. 19355
The American Rose Society	Old Jefferson Paige Road, Shreveport, Louisiana 71130
Bonsai Clubs International	c/o Francine Raphael, P.O. Box 2098, Sunnyvale, Calif. 94087
Bromeliad Society	Box 3279, Los Angeles, Calif. 90004
Cactus and Succulent Soc. of America	P.O. Box 3010, Santa Barbara, Calif. 93105
Cymbidium Society of America	1250 Orchard Drive, Santa Barbara, Calif. 93111
Delphinium Society	1630 Midwest Plaza Bldg., Minneapolis, Minn. 55402
Gesneriad Society International	P.O. Box 549, Knoxville, Tenn. 37901
The Herb Society of America	300 Massachusetts Avenue, Boston, Mass. 02115
Hobby Greenhouse Owners Ass'n.	45 Shady Drive, Wallingford, Conn. 06492
Holly Society of America	c/o Mr. B. Green Jr., 407 Fountain Green Road, Bel Air, Md. 21014
Hydroponics Society of America	P.O. Box 516, Brentwood, Calif. 94513
Indoor Light Gardening Society	35 Eton Road, Scarsdale, N.Y. 10583
International Cactus Society	P.O. Box 1452, San Angelo, Texas 76901
International Geranium Society	Dept. A.H.S., 6501 Yosemite Drive, Buena Park, Calif. 90620
National Chrysanthemum Society	c/o B. L. Markham, 2612 Beverly Blvd., Roanoke, Virginia 24015
North American Gladiolus Council	c/o Bob Dorsam, 30 Highland Place, Peru, Indiana 46970
North American Lily Society	c/o Earl Holl, P.O. Box 40134, Indianapolis, Indiana 46240
Palm Society	1320 S. Venetian Way, Miami, Fla. 33139
Saintpaulia International	P.O. Box 549, Knoxville, Tenn. 37901
Terrarium Association	57 Wolfpit Avenue, Norwalk, Conn. 06851

Ecology Related Organizations

Individuals interested in their environment and the conservation of our natural resources might consider membership in or request information from one of the following ecology-related organizations.

American Forestry Association
919 17th Street, N.W.
Washington, D.C. 20006

Elm Research Institute
Harrisville, N.H. 03450

Friends of the Earth
30 East 42nd Street
New York, N.Y. 10017

Izaak Walton League of America
1326 Waukegon Road
Glenview, Ill. 60025

National Audubon Society
1130 Fifth Avenue
New York, N.Y. 10028

National Parks and Conservation Society
1701 18th Street, N.W.
Washington, D.C. 20009

National Wildlife Federation
1412 16th Street, N.W.
Washington, D.C. 20006

The Nature Conservancy
Suite 800, 1800 North Kent Street
Arlington, Va. 22209

Sierra Club
1050 Mills Tower
San Francisco, Calif. 94104

The Wilderness Society
729 15th Street, N.W.
Washington, D.C. 20005

Appendix F

Addresses of Publication Suppliers

For those interested in obtaining additional information about a specific subject area, the following pamphlets and books are presented in the same sequence as the chapters in the textbook.

Single copies of many of the pamphlets listed may be obtained free by simply writing the supplier. Those pamphlets for which there is a slight charge (less than two dollars) will be designated with an asterisk.

The following is the list of publication suppliers for the pamphlets mentioned on the following pages.

U.S.D.A.
Office of Information
Washington, D.C. 20250

Mailing Room
Building 7
Research Park
Cornell University
Ithaca, N.Y. 14850

Scotts, The Lawn People
Marysville, Ohio 43040

American Society of Horticulture Science
P.O. Box 109
St. Joseph, Mich. 49085

Society of American Florists
901 North Washington Street
Alexandria, Va. 22314

The Stanford Seed Company
P.O. Box 366
Buffalo, N.Y. 14240

Brooklyn Botanic Garden
1000 Washington Avenue
Brooklyn, N.Y. 11225

University of Illinois, College of Agriculture
Vocational Agriculture Service
Urbana, Ill. 61801

Jackson and Perkins Company
Medford, Oreg. 97501

American Association of Nurserymen Inc.
230 Southern Building
Washington, D.C. 20005

George W. Park Seed Company
Greenwood, S.C. 29646

Florists' Transworld Delivery Association
900 West Lafayette
Detroit, Mich. 48226

Agway Incorporated
P.O. Box 1333
Syracuse, N.Y. 13201

Appendix G
Bibliography: Pamphlets and Books

Pamphlets

INTRODUCTION

Careers in the Floral Industry (Soc. of Amer. Florists)
Horticulture—A Rewarding Career (Amer. Soc. Hort. Science)
Develop a Career as a Professional Landscape Expert (Amer. Assoc. of Nurserymen)

Books

BASIC BOTANICAL BACKGROUND

Botany for Gardeners: Rickett (Macmillan)
Introductory Plant Science: Northen (Ronald)

Pamphlets

SOILS

Land Smoothing and Surface Drainage, No. E-1214 (Cornell)
Drainage Around the Home, No. IB-14 (Cornell)
Soils, No. 20 (Brooklyn Bot. Gar.)
Know Your Soil (U.S.D.A.)
Our American Land, Use the Land, Save the Soil (U.S.D.A.)
How To Control a Gully (U.S.D.A.)
The Effect of Soils and Fertilizers on Human and Animal Nutrition (U.S.D.A.)
Soil Surveys Can Help You (U.S.D.A.)
Maintaining Subsurface Drains (U.S.D.A.)

Books

A Book About Soils for the Home Gardener: Ortloff (Morrow)
Soils that Support Us: Kellogg (Macmillan)
Your Garden Soil: How To Make the Most Of It: Carleton (Van Nostrand)
The Soil and its Fertility: Teuscher (Van Nostrand-Reinhold)
Soils: An Introduction to Soils and Plant Growth: Donahue (Prentice-Hall)
Earth, the Stuff of Life: Bear (Univ. of Oklahoma Press)

CULTURAL
REQUIREMENTS

Pamphlets

Selecting Fertilizers for Lawns and Gardens, No. G-89 (U.S.D.A.)
*Your Tree's Troubles May Be You, No. 346C (U.S.D.A.)
*Fertilizing and Watering Shade and Ornamental Trees, No. 5003 (Univ. of Ill.)

CONTAINERS

Pamphlets

Container Culture of Ornamental Plants for the Home Grounds, No. E-1152 (Cornell)
*The Cornell Automated Plant Grower, No. IB-40 (Cornell)
*Gardening in Containers, No. 26 (Brooklyn Bot. Gar.)

Books

Container Gardening, Indoors and Out: Kramer (Doubleday)
Gardening in Containers: McDonald (Grosset and Dunlap)
Gardening in Containers: Sunset editors (Lane)
Hanging Gardens: Kramer (Scribner's)
Plant Containers: Sunset editors (Lane)
Planters—Make Your Own Containers: Kramer (Ballantine)

PROPAGATION

Pamphlets

Home Propagation of Ornamental Trees and Shrubs, No. G-80 (U.S.D.A.)
Flowers from Seed, No. IB-20 (Cornell)
Propagation of House Plants, No. E-1086 (Cornell)
New Shrubs From Old By Layering, No. E-1006 (Cornell)
*Propagation, No. 24 (Brooklyn Bot. Gar.)
*Producing Plants by Asexual Propagation, No. 5006 (Univ. of Ill.)

Books

Plant Propagation: Hartman (Prentice-Hall)
Plant Propagation Practices: Wells (Macmillan)
Plant Propagation in Pictures: Free (Doubleday)
Practicing Plant Parenthood: Baylis (101 Productions)
Simple Propagation: Prockter (Scribner's)
Water, Light and Love: Milstein (Applewood Seed Co.)

PLANTING

Pamphlets

Transplanting Ornamental Trees and Shrubs, No. G-192 (U.S.D.A.)
Suggested Practices for Planting and Maintaining Ornamental Trees and Shrubs on the Home Grounds, No. IB-24 (Cornell)
*Transplanting Shade Trees No. 5002 (Univ. of Ill.)

Lawn and Garden Guide (Agway)
Planting in Urban Soils, No. 345C (U.S.D.A.)

PRUNING

Pamphlets

Pruning Shade Trees and Repairing Their Injuries, No. G-83 (U.S.D.A.)
Pruning Ornamental Shrubs and Vines, No. G-165 (U.S.D.A.)
Illustrated Guide to Pruning Ornamental Trees and Shrubs on the Home Grounds, No. IB-23 (Cornell)
Pruning, No. 28 (Brooklyn Botanic Gar.)
Pruning Shade Trees, No. 5004 (Univ. of Ill.)

Books

How To Prune Almost Everything: Baumgardt (Barrows)
Pruning Handbook: Sunset editors (Lane)
Pruning Manual: Steffek (Van Nostrand-Reinhold)
The Complete Handbook of Pruning: Grounds (Macmillan)

PEST CONTROL

Pamphlets

Ants in the Home and Garden, No. G-28 (U.S.D.A.)
Insects and Diseases of Vegetables in the Home Garden, No. G-46 (U.S.D.A.)
Insects on Deciduous Fruits and Tree Nuts in the Home Orchard, No. G-190 (U.S.D.A.)
Safe Use of Pesticides in the Home and Garden, No. PA-589 (U.S.D.A.)
Poison Ivy, Poison Oak and Poison Sumac, No. F-1972 (U.S.D.A.)
Control of Grape Diseases and Insects in the Eastern United States, No. F-1893 (U.S.D.A.)
Controlling Diseases of Raspberries and Blackberries, No. F-2208 (U.S.D.A.)
Disease and Insect Control in the Home Orchard, No. 1082 (Cornell)
Precautions and First Aid Measures for Use in the Handling and Applying of Pesticides (Cornell)
Common Insects of Vegetables, No. 1035 (Cornell)
Identification and Control of Tomato Diseases in the Home Garden, No. 1076 (Cornell)
Pesticides and You, No. 6 (Cornell)
Controlling Undesirable Woody Plants with Chemicals, No. 1147 (Cornell)
Recommendations for Pest Control for Commercial Production and Maintenance of Trees and Shrubs (Cornell)
Controlling Household Pests, No. 202D (U.S.D.A.)
Tree and Shrub Insects and Their Control, No. 5005a (U. of Ill.)

*Controlling Insects on Fruit and Nut Trees, No. 312D (U.S.D.A.)
Keeping the Bad Bugs Out Naturally, No. 205D (U.S.D.A.)
Safe Use of Pesticides Around the Home, No. 208C (U.S.D.A.)
*Handbook on Garden Pests (Brooklyn Bot. Gar.)

Books
Diseases and Pests of Ornamental Plants: Pirone (Ronald)
Insects that Feed on Trees and Shrubs: Johnson (Cornell U. Press)
Plant Disease Handbook: Westcott (Van Nostrand-Reinhold)
The Gardener's Bug Book: Westcott (Doubleday)

LANDSCAPE DESIGN

Pamphlets
Home Planting by Design, No. G-164 (U.S.D.A.)
Landscape Design for Residential Property, No. E-1099 (Cornell)
*Homescaping (kit from U. of Wyoming, Laramie, Wyo. 82070)
*Landscaping Your Home, No. 858 (Univ. of Illinois)

Books
Art of Home Landscaping: Eckbo (McGraw-Hill)
Do's and Don't's of Home Landscape Design: Stoffel (Hearthside)
Home Landscaping You Can Design Yourself: Roberts (Hawthorn)
How To Plan Your Own Home Landscape: Weber (Bobbs-Merrill)
How To Plan Your Own Property: Ireys (Morrow)
Informal Gardening: Kilvert (Macmillan)
Landscaping and the Small Garden: Dietz (Doubleday)
Landscape Gardening: Crockett (Time-Life)
Room Outside: Brookes (Viking Press)

CONSTRUCTIONAL FEATURES

Pamphlets
*Building Hobby Greenhouses, No. 344C (U.S.D.A.)

Books
How To Build Walks, Walls and Patio Floors: Sunset editors (Lane)
Patios, Terraces, Decks and Roof Gardens: Smith (Hawthorn)
The Outdoor How-To-Build-It Book: Behme (Hawthorn)

LAWNS

Pamphlets
Better Lawns, Establishment, Maintenance, Renovation, Lawn Problems, Grasses, No. G-51 (U.S.D.A.)
Lawn Insects—How To Control Them, No. G-53 (U.S.D.A.)
Lawn Diseases—How To Control Them, No. G-61 (U.S.D.A.)
Lawn Weed Control With Herbicides, No. G-123 (U.S.D.A.)
Old English Lawn Handbook (Stanford Seed Company)
Recommendations for Turfgrass (Cornell)

Home Lawns, No. E-922 (Cornell)
What's That Weed? (Scotts)
On Watering Lawns, No. 14 (Scotts)
How To Succeed With Seed, No. 15 (Scotts)
Seasonal Guide for Established Lawns, No. 17 (Scotts)
Hot Dry Weather May Bring Undesirable Lawn Visitors, No. 23
   (Scotts)
Problems of Shade, No. 24 (Scotts)
On Mowing Lawns, No. 25 (Scotts)
Care of a Lawn Containing Bluegrass, No. 26 (Scotts)
*Better Lawns, No. 204C (U.S.D.A.)
*Lawn Weed Control, No. 281C (U.S.D.A.)

Books
Home Guide to Lawns and Landscaping: Cassidy (Harper & Row)
Lawn Book: Schery (Macmillan)
Lawn Keeping: Schery (Prentice-Hall)
Your Lawn: How To Make and Keep It: Carleton (Van Nostrand-
   Reinhold)

Pamphlets
Let's Plant a Tree, No. 278D (U.S.D.A.)
Trees for Shade and Beauty, No. G-117 (U.S.D.A.)
Shrubs, Vines, and Trees for Summer Color, No. G-181 (U.S.D.A.)
Selecting Shrubs for Shady Areas, No. G-142 (U.S.D.A.)
Color It Green With Trees, No. PA-791 (U.S.D.A.)
Shaping Christmas Trees for Quality, No. E-1080 (Cornell)
Forcing Shrubs and Trees for Indoor Bloom, No. E-1030 (Cornell)
Growing Trees in Small Nurseries, No. E-1198 (Cornell)
Home Garden Guide For Ornamental Trees and Shrubs (Cornell)
*100 Finest Trees and Shrubs, No. 25 (Brooklyn Bot. Gar.)
*Flowering Trees, No. 41 (Brooklyn Botanic Garden)

Books
A Field Guide to Trees and Shrubs: Petrides (Houghton Mifflin)
Flowering Trees and Shrubs in Color: Mondino (Doubleday)
The Gardener's Basic Book of Trees and Shrubs: Schuler (Simon and
   Schuster)
Trees: Crockett (Time-Life)
Trees for American Gardens: Wyman (Macmillan)

Books
How To Build Fences and Gates: Sunset editors (Lane)

SPECIMEN TREES

FENCES AND
HEDGES

MULCHES AND
GROUND COVERS

Pamphlets

Growing Ground Covers, No. G-175 (U.S.D.A.)
Mulches for Your Garden, No. G-185 (U.S.D.A.)
Ground Covers in New York State, No. E-1178 (Cornell)
Ornamental Vines for Landscape Plantings in New York State, No. E-1122 (Cornell)
Plastic Mulches, No. E-1180 (Cornell)
Mulches, No. 23 (Brooklyn Botanic Garden)

Books

Complete Book of Ground Covers: Atkinson (McKay)
Ground Cover Plants: Wyman (Macmillan)

GARDENS

Pamphlets

Growing Flowering Perennials, No. 210C (U.S.D.A.)
Introduction to Home Gardening, No. E-1049 (Cornell)
The Gardener's Handbook (George W. Park Seed Co.)

Books

America's Garden Book: Bush-Brown (Scribner's)
Better Homes and Gardens New Garden Book: Meredith
Gardening for Beginners: Hutchison (Collier)
Herbs for Every Garden: Foster (Dutton)
Home Gardening At its Best: Gilbertie (Atheneum)
McCall's Garden Book: Harshbarger (Simon and Schuster)
N.Y. Times Garden Book: Faust (Ballantine)
Small Garden Book: Carleton (Macmillan)
Small Gardens Are More Fun: Tarantino (Simon and Schuster)
Successful Gardening With Perennials: Wilson (Doubleday)
The Complete Book for Gardeners: Snyder (Van Nostrand)
The Whole Seed Catalog: Link and Stark (Madison Square)

VEGETABLES

Pamphlets

Growing Tomatoes in the Home Garden, No. G-180 (U.S.D.A.)
Minigardens for Vegetables, No. G-163 (U.S.D.A.)
Vegetable Production Recommendations (Cornell)
The Home Vegetable Garden, No. 69 (Brooklyn Bot. Gar.)

Books

Crockett's Victory Garden: Crockett (Little, Brown)
Farming for Self-Sufficiency: Seymour (Schocken)
From Garden to Table: Fielden (McClelland and Stewart)
How To Eat Better and Spend Less: A Complete Guide to Vegetable Gardening: Rice (Reston)
N.Y. Times Book of Vegetable Gardening: Faust (A&W)

Small World Vegetable Gardening: Bryan (Scribner's)
The Complete Vegetable Gardener: Seabrook (A&W)
Vegetable Gardening: Sunset editors (Lane)
Vegetable Gardening and Cooking: Hoobler (Grosset and Dunlap)

Pamphlets
Why Fruit Trees Fail To Bear, No. L-172 (U.S.D.A.)
Dwarf Fruit Trees—Selection and Care, No. L-407 (U.S.D.A.)
Strawberry Culture, No. F-1028 (U.S.D.A.)
Growing Blackberries, No. F-2160 (U.S.D.A.)
Growing Raspberries, No. F-2165 (U.S.D.A.)
Growing Ginseng, No. F-2201 (U.S.D.A.)
Establishing and Managing Young Apple Orchards, No. F-1897
    (U.S.D.A.)
Tree Fruit Production Recommendations (Cornell)
*Fruit Trees and Shrubs, No. 67 (Brooklyn Botanic Gar.)

Books
Dwarf Fruit Trees, Indoors and Outdoors: Atkinson (Van Nostrand-
    Reinhold)
Fruits and Berries for the Home Garden: Hill (Knopf)
Small Fruit Culture: Shoemaker (McGraw-Hill)
The Green Thumb Book of Fruit and Vegetable Gardening: Abraham
    (Prentice-Hall)

Pamphlets
Selecting and Growing House Plants, No. G-82 (U.S.D.A.)
Care and Use of Plants in the Home, No. E-1073 (Cornell)
Care of Flowering Plants in the Home, No. E-1117 (Cornell)
*Insects and Related Pests of House Plants, and How To Control
    Them,* No. G-67 (U.S.D.A.)
*House Plants, No. 40 (Brooklyn Bot. Gar.)
*House Plant Primer, No. 70 (Brooklyn Bot. Gar.)
*Growing Plants Indoors, No. 5007 (Univ. of Illinois)

Books
All About House Plants: Free (Doubleday)
Exotic Plant Manual: Graf (Scribner's)
Flowering House Plants: Crockett (Time-Life)
Flowering House Plants Month by Month: Kramer (Cornerstone)
Foliage House Plants: Crockett (Time-Life)
Fun with Growing Herbs Indoors: Elbert (Crown)
Fun with Growing Odd and Curious House Plants: Elbert (Crown)
Garden in Your House: Ballard (Harper and Row)
House Plants Are for Pleasure: Wilson (Doubleday)

House Plants for Five Exposures: Taloumis (N.A.L.)
How To Grow House Plants: Sunset editors (Lane)
Making Things Grow Indoors: Cruso (Knopf)
Plants that Really Bloom Indoors: Elbert (Simon and Schuster)
The Complete Book of House Plants: Fitch (Hawthorn)
The Complete Indoor Gardener: Wright (Random House)
1000 Beautiful House Plants and How To Grow Them: Kramer (Morrow)

INDOOR
LIGHTING

Pamphlets
Artificial Lighting for Decorative Plants, No. E-1087 (Cornell)
Indoor Gardens with Controlled Lighting, No. G-187 (U.S.D.A.)

Books
Fluorescent Light Gardening: Cherry (Van Nostrand-Reinhold)
Gardening Indoors Under Lights: Kranz (Viking Press)
How To Grow Beautiful Flowers, Vegetables and Plants Indoors With Lights: Erikson (Duro-Lite Lamps)
The Complete Book of House Plants Under Lights: Fitch (Hawthorn)
The Indoor Light Gardening Book: Elbert (Crown)

FLOWER
ARRANGING

Pamphlets
Flowers and You (Society of American Florists)
Corsages from Garden Flowers, No. E-1047 (Cornell)
Be an Artist: Make Your Own Pressed Flower Pictures, No. IB-34 (Cornell)
Christmas Decorations, No. E-379 (Cornell)
Starting and Managing a Retail Flower Shop (F.T.D.A.)
Flower Arrangement, No. 19 (Brooklyn Bot. Gar.)
Arranging Flowers in Vases and Bowls, No. 5009 (U. of Illinois)

Books
ABC of Flower Arranging: Clements (N.Y. Times Book Co.)
Better Homes and Gardens Flower Arranging (Meredith)
Design for Flower Arrangers: Riester (Van Nostrand)

GREENHOUSE
MANAGEMENT

Pamphlets
Recommendations for Commercial Floriculture Crops (Cornell)
Plastic Covered Greenhouses, No. 72 (Cornell)
Home Greenhouse and Gardening Structures, No. 1043 (Cornell)
Greenhouse—Plastic Covered Plan #6094, No. M-1202 (U.S.D.A.)
Lath House for Nursery Plants Plan #6064, No. M-1154 (U.S.D.A.)

Books
Gardening Under Glass: Eaton (Macmillan)
Greenhouse Gardening: Northern (Ronald)
The Greenhouse Catalog of Catalogs: Lapidus (McKay)
The Greenhouse Gardener: McDonald (New American Library)
Your Window Greenhouse: Kramer (Crowell)

Pamphlets
Facts about Organic Gardening, No. IB-36 (Cornell)
Guide to Natural Gardening (Agway)
The Compost Pile, No. E-991 (Cornell)

*COMPOST PILE
AND ORGANIC
GARDENING*

Books
Growing Food the Natural Way: Kraft (Doubleday)
Guide to Organic Gardening: Sunset editors (Lane)
How To Farm Your Backyard the Mulch-Organic Way: Alth
  (McGraw-Hill)
Organic Gardening Without Poisons: Tyler (Van Nostrand)
Organic Vegetable Growing: Ogden (Rodale Press)

Pamphlets
Bulbous Plants for Indoor Bloom, No. E-1021 (Cornell)
Spring Flowering Bulbs, No. 340D (U.S.D.A.)

FORCING BULBS

Books
Bulbs: Crockett (Time-Life)
Complete Book of Garden Bulbs: Reynolds (Funk and Wagnalls)
Flowering Bulbs for Winter Windows: Walker (Van Nostrand)
The Wonderful World of Bulbs: Miles (Van Nostrand)

Pamphlets
Christmas Decorations Made With Plant Material, No. 344D
  (U.S.D.A.)

*CONDITIONING
PLANT MATERIAL*

Books
Complete Book of Flower Preservation: Condon (Prentice-Hall)
Dried Flowers: Yarburgh-Bateson (Scribner's)
Keeping the Plants You Pick: Foster (Crowell)

Pamphlets
Make Your Own Terrarium, No. 207C (U.S.D.A.)
The Terrarium, Series 4, No. 1 (Audubon Nature Bulletin)

TERRARIUMS

286

Bibliography

Books

All About Miniature Plants and Gardens: Brilmayer (Doubleday)
Bottle Gardens and Fern Cases: Ashberry (London, Hodder, Stoughton)
Gardening in Miniature: Genders (Robert Hale Ltd.)
Gardens in Glass Containers: Baur (Hearthside)
Gardens Under Glass: Kramer (Simon and Schuster)
Miniature Plants: McDonald (Van Nostrand)
Successful Terrariums: Kanyatta (Houghton Mifflin)
Terrariums and Miniature Gardens: Sunset editors (Lane)

CHRYSAN- **Books**
THEMUMS
A.B.C. of Chrysanthemums: Shewell-Cooper (Intl. Pub. Serv.)
Chrysanthemum Book: Cumming (Van Nostrand-Reinhold)
Chrysanthemums the Year Round: Searle (Intl. Pub. Serv.)

BEGONIAS **Books**

All About Begonias: Brilmayer (Doubleday)
Begonias: Indoors and Out: Kramer (Dutton)
Tuberous Begonias: Langdon (Intl. Pub. Serv.)

ROSES **Pamphlets**

Home Garden Guide (Jackson and Perkins)
*Roses for the Home, No. 350C (U.S.D.A.)

Books

Anyone Can Grow Roses: Westcott (Van Nostrand-Reinhold)
Complete Book of Miniature Roses: Fitch (Hawthorn)
Growing Better Roses: Nisbet (Knopf)
How To Grow Roses: Sunset editors (Lane)
Roses: Crockett (Time-Life)
The Rockwell's Complete Book of Roses: Rockwell (Doubleday)

FERNS **Pamphlets**

*Ferns (Brooklyn Botanic Garden)

Books

Ferns To Know and Grow: Foster (Hawthorn)

VINES **Books**

Climbing Plants: Bartrum (Branford)

Books
All About Geraniums: Schulz (Doubleday)
Geraniums and Pelargoniums: Fogg (Branford)
Joy of Geraniums: Wilson (Morrow)

Books
African Violets, Gloxinias and Their Relatives: Moore (Macmillan)
All About African Violets: Free (Doubleday)
Begonias, Gloxinias, and African Violets: Fogg (Branford)
Easy Guide to African Violets: Meachem (Hearthside)
Gloxinias and How To Grow Them: Schulz (Morrow)
How To Grow African Violets: Sunset editors (Lane)
The Complete Book of African Violets: Wilson (Burrows)

Books
Cacti: Borg (Intl. Pub. Serv.)
Cacti and Other Succulents: Chidamian (Doubleday)
Cacti and Succulents: Haage (Dutton)
Cacti and Succulents: Van Ness (Van Nostrand-Reinhold)
Succulents and Cactus: Sunset editors (Lane)

Pamphlets
Growing Bonsai, No. 206C (U.S.D.A.)
Dwarfed Potted Trees—The Bonsai of Japan (Brooklyn Bot. Gar.)

Books
Bonsai: Sunset editors (Lane)
Japanese Art of Miniature Trees and Landscapes: Yoshimura (Tuttle)

Appendix H

Artificial Soil Mixture and pH Scale

INGREDIENTS Ingredients and procedure for making an artificial soil mix

1. ½ bushel of peat moss
2. ½ bushel of vermiculite
3. 6 tablespoons of ground limestone
4. 8 to 10 tablespoons of 5-10-5 fertilizer

PROCEDURE

1. Mix the peat moss and vermiculite on a clean surface and moisten the mixture slightly.
2. Allow time for the peat moss to become moist before mixing in the ground limestone and fertilizer.
3. Once all the ingredients are thoroughly mixed, the artificial soil mix is ready to be used. Any excess can be stored indefinitely in any airtight container.

Note. This mix can be reused once it has been dried thoroughly and the roots from previous plants have been removed. Because the limestone and fertilizer have been removed by the previous plants, it is advisable to add both to the mixture and mix thoroughly before reusing.

THE pH SCALE

Soil condition	pH value
Medium alkaline	above 7.6
Slightly alkaline	7.1 to 7.5
Neutral	7.0

THE pH SCALE

Soil condition	pH value
Slightly acid	6.5 to 7.0
Medium acid	5.6 to 6.5
Strongly acid	below 5.5

Note. pH is a logarithmic scale where a single degree of difference, i.e., 5.5 to 6.5, represents a ten-times difference in acidity. Even the slightest change in pH may have an extreme effect on plant growth, bacterial activity, and nutrient availability.

Soil pH Preference

pH 5 to 5.5	pH 5.5 to 6.5	pH 6.5 to 7.0	
Bean	Bean	Asparagus	*FOR VEGETABLES*
Corn	Brussels sprouts	Beets	
Cucumber	Carrot	Broccoli	
Endive	Cabbage	Cauliflower	
Parsley	Corn	Celery	
Pea	Cucumber	Chard	
Peanut	Parsley	Eggplant	
Potato	Pea	Kale	
Pumpkin	Pepper	Lettuce	
Radish	Pumpkin	Muskmelon	
Sweet potato	Radish	Onion	
Tobacco	Squash	Parsnip	
Turnip	Sweet potato	Spinach	
Watermelon	Tobacco		
	Tomato		
	Turnip		

pH 5 to 5.5	pH 5.5 to 6.5	pH 6.5 to 7.0	
Blackberry	Blackberry	Apple	*FOR FRUITS*
Blueberry	Rhubarb	Cherry	
Raspberry	Strawberry	Currant	
Strawberry		Gooseberry	
		Grape	
		Peach	
		Pear	
		Plum	

pH 5 to 5.5	pH 5.5 to 6.5	pH 6.5 to 7.0	
Bentgrass	Clover	Bluegrass	*FOR LAWN GRASSES*
Orchard grass	Orchard grass	Clover	
Red top	Red top	Fescues	
Rye	Rye		

Metric Conversion Table

Pound	Ounce	Gram	Milligram
1	16		
¹/₁₆	1	ⓐ 28	ⓐ 28,000
	¹/₆	5	5,000
	¹/₃₀	1	1,000
	¹/₆₀	0.5	500
	¹/₁₀₀	0.3	300

METRIC
EQUIVALENTS

1 acre = 0.404 hectare

1 square foot = 0.092 square meter

1 gallon = 3.785 liters = 3785 milliliters

1 cubic foot = 0.028 cubic meter

1 pound = 0.454 kilogram = 454 grams

1 inch = 2.54 centimeters

1 foot = 0.305 meter = 30.5 centimeters

1 pint = 0.47 liter = 470 milliliters

32°F = 0°C (a change of 9°F = a change of 5°C)

1 bushel =2150.42 cubic inches

1 cubic inch = 16.387 cubic centimeters

Maintenance of Power Tools

To prolong the life of your gasoline-powered lawn mower, the following procedures should be followed:

1. Replace the spark plug every year.
2. Change the crankcase oil every 30 hours of running time.
3. Use high test gasoline.
4. Remove the grass clippings from the bottom of a rotary mower after each mowing. Make certain to disconnect the spark plug wire during the entire operation.
5. Clean the air cleaner at least once a month.
6. Drain the gas supply completely for winter storage.
7. Keep the blade on a rotary mower sharp and tight. (When checking the blade, disconnect the spark plug wire.)

Other gasoline-powered equipment includes the chain saw, the garden tractor, and the cultivator or rototiller.

Electric-powered equipment, such as grass and hedge clippers, chain saws, and lawn mowers, are preferred by some homeowners over their gasoline counterparts. The reasons usually cited include no fuel to add, no starting difficulties, and no spark plug to replace. The maintenance requirement of most electric power tools involves the addition of a small quantity of machine oil to the motor at prescribed intervals of time. The homeowner's biggest problem with electric-powered tools is to avoid cutting the electric cord in half.

Appendix L

Selection of a Garden Hose

There are several factors that must be considered prior to the purchase of a garden hose. They are

1. *The construction material.* In most cases, a *rubber* hose is more expensive than a *plastic* one simply because it is more durable.

2. *How the hose is constructed.* A garden hose can be either thin-walled or reinforced. For jobs requiring only low water pressure, the thin-walled hose is sufficient.

3. *The inside diameter.* A larger-diameter hose gives a larger volume of water with less pressure because of a loss due to friction. For example, a $3/4$-in.-(1.9-cm)-diameter hose delivers three to five times the volume that a $3/8$-in.-(0.95-cm)-diameter hose does. The length of the hose is also a factor. A shorter hose will deliver more water than a longer one.

4. *Attachments and their use.* What types of plants will be watered is the greatest consideration for the selection of these attachments.

5. *Maintenance.* All garden hoses must be drained and placed in storage for the winter season. Remember, also, to shut off and drain all outdoor faucets to prevent the pipes from rupturing due to the formation of ice.

African violet	Pilea	*Leaf Cuttings*
Aloe (burn plant)	Moses on a raft	
Croton	Rex begonia	
Goxinia	Sedum	
Gynura (purple passion)	Semi-tuberous begonia	
Kalanchoe	(Christmas)	
Peperomia	Snake plant (Sansevieria)	
	Wax plant (Hoya)	
Aralia	Geranium	*Stem Cuttings*
Azalea	Impatiens	
Begonia	Ivy—all types	
Camellia	Philodendrons—all types	
Christmas cactus	Pilea	
Chrysanthemum	Pothos	
Coleus	Poinsettia	
Fittonia	Shrimp plant	
Fuchsia	Tradescantia	
Gardenia		
African violet	Cyclamen	*From Seed*
Asparagus fern	Episcia	
Begonia—wax	Ficus	
Browallia	Fuchsia	
Cactus	Gloxinia	
Calceolaria	Impatiens	
Cinneria	Ornamental poppy	
Coleus	Jerusalem cherry	
Crossandra	Sensitive plant	
Corn plant (dracena)		*Air Layering*

Dumb cane (Dieffenbachia)
Gardenia
Rubber plant (Ficus)

New Plants Form   On Leaves   Pick a back plant (Tolmiea)

Tubers   Anemone
Begonia, tuberous
Caladium
Gloxinia
Ranunculus
Rechsteineria
Smithiantha

Division   African violet
Baby tears
Chinese evergreen
Ferns
Orchids
Parlor plant (Aspidistra)
Strawberry geranium

Runners   Bromeliad

Pups   **Bromeliad**

Proper Use and Handling of Pesticides

1. The directions and special precautions that are printed on the label of each pesticide must be followed.
2. When not being used, all pesticides should be locked in a well-ventilated cabinet out of the reach of both children and pets.
3. Every attempt should be made to keep the labels on each pesticide container from becoming illegible. Whenever a container is finally emptied, it should be properly disposed of.
4. No pesticide should be opened, mixed, or applied without adequate ventilation.
5. The applicator should avoid the inhalation of any pesticide particles and avoid letting it touch either his skin or clothing. The use of a respirator is advisable at all times. Whenever contact to the skin or clothing occurs, immediately bathe the area with soap and water.
6. Pesticides are best applied when the plant material is dry, so that good coverage is insured, and when the air is still, so that the pesticide does not drift into the wrong area.
7. If it rains within 24 hours of any pesticide application, a second application should be made as soon as the foliage has become completely dry.
8. It is important to realize that the effectiveness of any pesticide application is based on:
   a. the time of application
   b. the form in which the material is applied—i.e., dust or spray
   c. the concentration used
   d. the amount that runs off
   e. the amount that is leached away
   f. the temperature
   g. the exposure to sunlight

Directions for Mixing Your Own Pesticide

1. Use only wettable powders.
2. Mix only what you intend to use.
3. Apply immediately.
4. Shake the mixture frequently to help keep the powder in suspension.
5. Repeat the application, when necessary, every two weeks.

Plant Damage due to Gas Poisoning and Air Pollution

Gas Poisoning   Due to escaping or leaking gas that is not completely combusted.

Symptoms.   A rapid yellowing and dropping of leaves shortly after exposure to the gas.

Test plants.   Tomato plants whose leaves droop sharply after exposure and freshly cut carnations whose petals curl inward rapidly.
    House plants *not* harmed by small quantities of gas:

Amaryllis	Holly fern
Anthurium	Hoya
Begonia	Impatiens
Bromeliad	Philodendron
Cactus	Poinsettia
Dracena	Scindapsus
English ivy	Tolmeia
Ficus	Tradescantia
Gardenia	

Air Pollution   **Smog.**   General term used to describe all the gaseous pollutants in the atmosphere.

Photochemical smog.   Smog acted upon chemically by sunlight.

Gaseous pollutants.   Ammonia, chlorine, hydrogen chloride, hydrogen fluoride, hydrogen sulfide, nitrogen dioxide, ozone, sulfur dioxide, and mercury vapor.
    Plant characteristics that provide some resistance to smog are

1. Small leaves
2. Slow growth habit
3. Small flowers

Buxus

Camellia

Cotoneaster

Cupressus (cypress)

Fraxinus

Ginkgo

Prunus

Pyracantha

Quercus

Spiraea

Syringa

Viburnum

Note. To help you confirm plant damage due to air pollution, a color guide showing specific damage can be purchased from the U.S. Government Printing Office, Washington, D.C. 20402 for $1.25. The publication is entitled *National Air Pollution Control Administration Publication #* AP-71.

Appendix P

Suggested Ingredients for a Finished Landscape

1. That all lawn areas be
   a. Planted with the correct grass type
   b. Well drained
   c. Properly graded
   d. Neat in appearance
   e. Weed-free
2. That the following types of plantings all be well designed:
   a. Foundation
   b. Gardens. Examples include:
      (1) Flower—annuals, perennials, bulbs
      (2) Vegetable
      (3) Aquatic
      (4) Rose
      (5) Rock or Alpine
      (6) Herb
   c. Ground covers
   d. Hedges
   e. Shrubs
   f. Trees
3. That the following constructional features be well positioned and blend well with the total landscape picture:
   a. Animal structures—bird houses, feeders, and baths
   b. Benches
   c. Fences
   d. Garden houses
   e. Sculpture and statuary
   f. Steps
   g. Walks
   h. Walls
   i. Water structures—pools, fountains, and swimming pools

4. That the following private or intimate areas be both functional and pleasing to the eye:

a. Barbecue area

b. Deck

c. Patio

d. Play area

e. Terrace

Appendix Q

Lawn Calendar

Month	Procedures to Follow
March	1. Fertilize all the lawn area (minimum 10-6-4). 2. Renovate all the bare areas. 3. Apply ground limestone if the soil is acidic. 4. Lawn areas can be top dressed with humus if the time and materials are available.
April	1. Apply pre-emergence crabgrass killer (Bensulide) on mature lawns. *Note:* The surface must be dry and recently mowed clippings removed. 2. Kentucky bluegrass should be sown before the 10th.
May	1. Fescues and ryegrass can be sown during this month. 2. Use the pre-emergence crabgrass killer (Siduron) on newly seeded areas through early June.
June	1. Raise the cutting edge of the lawn mower ½ in. 2. Apply 2,4-D in liquid form to destroy most broadleaf weeds. *Note:* Do not allow the spray to drift into planted areas other than the lawn.
July	1. For areas to be sown in the fall, sow buckwheat, oats, or ryegrass during the first two weeks. These plants should be turned under five to six weeks later as a source of organic matter before the actual lawn seeding occurs. 2. Continue to mow at a minimum height of 2 in.
August	1. Prepare the lawn areas for new seeding during the first half of the month by thoroughly turning into the soil the source of organic matter, plus the required amounts of lime and fertilizer.
September	1. Seed the new area within the first week to 10 days. 2. Make certain that the lawn seed bed never completely dries out.
November	1. Apply fertilizer to all lawn areas to ensure continued, yet much slower, growth during the winter months and a faster recovery in the spring.

300

Grass (Scientific name)	Planted by	Spreads by	Mowing height (in.)	Soil preference	Pest-free	Shade tolerant	Additional comments
Northern regions							
Carpet grass (Axonopus compressus)	seed sprigs		1 (2.5 cm)	high moisture poor soil	yes	no	Withstands heavy wear; needs sufficient iron
Canada bluegrass (Poa compressa)	seed		1½ (3.8 cm)	good drainage high pH, poor soil		no	Used in playground areas; it is tough and resists wear
Colonial bentgrass (Agrostis tenuis)	seed	creeping stems	close	good drainage high fertility	no	no	Used for golf putting greens
Kentucky bluegrass (Poa pratensis)	seed sod	root stocks	1½ (3.8 cm)	good drainage high pH high fertility	no	no	Backbone of all sunny lawns
Merion bluegrass (P. pratensis variety Merion)		consult Kentucky bluegrass			no	no	More drought resistant than Kentucky bluegrass
Red fescue (Festuca rubra)	seed	root stocks	above 1½ (3.8 cm)	can grow in poor soils with high acidity	no	yes	Slow-growing grass; can stand long periods of drought
Rough bluegrass (Poa trivialis)	seed	runners	above 1½ (3.8 cm)	high moisture	no	yes	Injured by hot, dry weather; used only in areas of no traffic, shallow roots
Tall fescue (Festuca arundinacea)	seed		above 1½ (3.8 cm)	prefers fertile, well-drained soil but can tolerate various soils		yes	Forms a tough turf and is used on playing fields; may be a nuisance in the lawn
Southern regions							
Bermuda grass (Cynodon dactylon)	seed sprigs	runners root stocks		good drainage high pH, high fertility	yes	no	Needs sun; destroyed by shade; high acidity or low fertility soils
Centipede grass (Eremochloa ephiruroides)	sprigs seed			best grass for low maintenance	yes		Vigorous grass; must have sufficient iron
Japanese lawngrass (Zoysia japonica)	sprigs	runners root stocks	close	able to grow in infertile soils	yes	no	Turns straw color in winter; resistant to drought and wear
St. Augustine grass (Stenotaphrum secundatum)	sprigs	runners		moist, fertile soils	no	yes	Needs liberal nitrogen
For temporary lawns							
Annual ryegrass (Lolium multiflorum)	seed	reseeding		cool, moist soil, good fertility			Germinates quickly, but usually dies at the end of the year
Perennial ryegrass (Lolium perenne)	seed	reseeding		same as for annual rye			Fast germinator, but never forms a thick sod; helps prevent erosion on steep banks
Red top (Agrostis alba)	seed		above 1½ (3.8 cm)	can grow in wide variety of soils	no	no	Quick to germinate; resists drought; becomes clumpy when it persists in the lawn

Appendix S

Trees For Special Purposes

Fast Growers	American elm	Scotch pine
(large)	Basswood	Sweet gum
	Catalpa	Sycamore
	European mountain ash	Tulip tree
	Honey locust	White pine
	London plane tree	
	Norway maple	(small) Crabapple
	Norway spruce	Dogwood
	Pin oak	Goldenrain tree
	Red maple	Silverbell tree
	Red oak	Washington thorn
	Scarlet oak	
For a Small	Austrian pine	Moraine locust
Space	Flowering crabapple	Redbud
	Flowering dogwood	Saucer magnolia
	English hawthorn	Silverbell tree
	European mountain ash	Sourwood
	Ginkgo	Star magnolia
	Goldenrain tree	Sweet gum
	Japanese flowering cherry	White birch
	Japanese tree lilac	
For Dry, Sandy	Chinese juniper	Red cedar
Soils	Hedge maple	Scotch pine
	Norway spruce	White pine
		White spruce
Street Shade Trees	American elm	London plane tree
	Basswood	Magnolia
	Catalpa	Norway maple
	English hawthorn*	Pin oak
	Ginkgo	Sugar maple
	Goldenrain tree*	Sweet gum
	Japanese tree lilac*	Washington thorn*
		*Narrow street

302

Cherry	Fringe tree	*Provide Food*
Crabapple	Hawthorn	*for the Birds*
Dogwood	Holly	
European mountain ash	Mulberry	
Douglas fir	Magnolia	*Free from Insects*
Goldenrain tree	Silverbell tree	*and Diseases*
Ginkgo	Sourgum	
Hemlock	Sourwood	
Katsura-tree	Sweet gum	
Arborvitae	Red maple	*For Wet Soils*
Balsam fir	Red spruce	
Basswood	Sourgum	
Hemlock	Sweet gum	
Pin oak	Weeping willow	
Arborvitae	Red cedar	*Used as*
Austrian pine	Red spruce	*Windbreaks*
Hemlock	Scotch pine	
Japanese black pine	White pine	
Norway spruce	White poplar	
Pin oak	White spruce	
Andromeda	Pine	*Acid-loving Trees*
Azalea	Red cedar	*and Shrubs*
Fir	Rhododendron	*(pH 5 to 5.5)*
Hemlock	Spruce	
Holly	White birch	
Mountain laurel	White cedar	
Red maple	Sourgum	*Able to Grow in*
Red oak	White poplar	*Windy Locations*
Sassafras	White willow	

Appendix T

Desirable Shade Trees For Specific Geographical Locations

Location Common name (scientific name)	Maximum height (ft)	Comments
South		
American holly (*Ilex opaca*)	40 (12.2 m)	Evergreen
Black olive (*Bucida buceras*)	40 (12.2 m)	Evergreen
Camphor tree (*Cinnamomum camphora*)	40 (12.2 m)	Evergreen
Glossy privet (*Ligustrum lucidum*)	25 (7.6 m)	Evergreen
Live oak (*Quercus virginiana*)	75 (23 m)	Evergreen
Loblolly pine (*Pinus taeda*)	100 (31 m)	Evergreen
Loquat (*Eribotrya japonica*)	25 (7.6 m)	Evergreen
Pongam (*Pongamia pinnata*)	40 (12.2 m)	Semi-evergreen
Sapodilla (*Achras zapota*)	40 (12.2 m)	Evergreen
Southern magnolia (*Magnolia grandiflora*)	60 (18.3 m)	Evergreen
Sugarberry hackberry (*Celtis laevigata*)	90 (27.5 m)	Deciduous
Southwest		
African sumac (*Rhus lancea*)	25 (7.6 m)	Evergreen
California pepper tree (*Schinus molle*)	40 (12.2 m)	Evergreen
Chinese elm (*Ulmus parvifolia*)	60 (18.3 m)	Semi-evergreen
Fruitless mulberry (*Morus alba*)	60 (18.3 m)	Deciduous
Japanese privet (*Ligustrum japonicum*)	12 (3.67 m)	Evergreen
Modesto ash (*Fraxinus velutina*)	50 (15.3 m)	Deciduous
Olive (*Olea europaea*)	35 (10.7 m)	Evergreen
Oriental plane tree (*Platanus orientalis*)	100 (31 m)	Deciduous
Red box gum (*Eucalyptus polyanthemos*)	100 (31 m)	Evergreen
St. Johns bread (*Ceratonia siliqua*)	40 (12.2 m)	Evergreen
Southern California		
Aleppo pine (*Pinus halepensis*)	60 (18.3 m)	Evergreen
Brazilian pepper tree (*Schinus terebinthifolius*)	30 (9.2 m)	Evergreen
Camphor tree (*Cinnamomum camphora*)	40 (12.2 m)	Evergreen
Chinese elm (*Ulmus parvifolia*)	60 (18.3 m)	Semi-evergreen
European hackberry (*Celtis australia*)	75 (23 m)	Deciduous
Fruitless mulberry (*Morus alba*)	60 (18.3 m)	Deciduous
Modesto ash (*Fraxinus velutina*)	50 (15.3 m)	Deciduous
St. Johns bread (*Ceratonia siliqua*)	40 (12.2 m)	Evergreen
Strawberry tree (*Arbutus unedo*)	20 (6.1 m)	Evergreen

Location Common name (scientific name)	Maximum height (ft)	Comments
Northern California		
Chinese elm (*Ulmus parvifolia*)	60 (18.3 m)	Semi-evergreen
Chinese pistachio (*Pistacia chinensis*)	60 (18.3 m)	Deciduous
Evergreen ash (*Fraxinus uhdei*)	30 (9.2 m)	Semi-evergreen
Fruitless mulberry (*Morus alba*)	60 (18.3 m)	Deciduous
Holly oak (*Quercus ilex*)	40 (12.2 m)	Evergreen
Japanese zelkova (*Zelkova serrata*)	80 (24.4 m)	Deciduous
Modesto ash (*Fraxinus velutina*)	50 (15.3 m)	Deciduous
Southern magnolia (*Magnolia grandiflora*)	60 (18.3 m)	Evergreen

Desirable Flowering Trees For Specific Geographical Locations

Location Common name (scientific name)	Maximum height (ft)	Flower color
South		
Bull bay (*Magnolia grandiflora*)	60 (18.3 m)	white
Callery pear (*Pyrus calleryana*)	30 (9.2 m)	white
Carolina silver bell (*Halesia carolina*)	30 (9.2 m)	white
Crape myrtle (*Lagerstroemia indica*)	20 (6.1 m)	pink, red, white
Jacaranda (*Jacaranda acutifolia*)	50 (15.3 m)	violet-blue
Jerusalem thorn (*Parkinsonia aculeata*)	30 (9.2 m)	yellow
Orchid tree (*Bauhinia variegata*)	20 (6.1 m)	purple
Royal poinciana (*Delonix regia*)	40 (12.2 m)	scarlet, orange
Shower of gold (*Cassia fistula*)	30 (9.2 m)	yellow
Sweet bay magnolia (*Magnolia virginiana*)	30 (9.2 m)	white
Southwest		
Blue paloverde (*Cercidium torreyanum*)	20 (6.1 m)	yellow
Chaste-tree (*Vitex agnus-castus*)	10 (3.1 m)	lavender-blue
Cootamundra Wattle (*Acacia baileyana*)	30 (9.2 m)	yellow
Crape myrtle (*Lagerstroemia indica*)	20 (6.1 m)	pink, red, white
Pomegranate (*Punica granatum*)	12 (3.7 m)	yellow, orange
Southern magnolia (*Magnolia grandiflora*)	60 (18.3 m)	white
Southern California		
Coral tree (*Erythrina*)	40 (12.2 m)	red
Flame eucalyptus (*Eucalyptus ficifolia*)	35 (10.7 m)	red, salmon, white
Floss-silk tree (*Chorisia speciosa*)	50 (15.3 m)	yellow, purple, red, pk.
Jacaranda (*Jacaranda acutifolia*)	50 (15.3 m)	violet-blue
Mexican paloverde (*Parkinsonia aculeata*)	30 (9.2 m)	yellow
Red bottle brush (*Callistemon citrinus*)	25 (7.6 m)	red
Sweetshade (*Hymenosporum flavum*)	40 (12.2 m)	yellow
Southern magnolia (*Magnolia grandiflora*)	60 (18.3 m)	white
Tipu-tree (*Tipuana tipu*)	60 (18.3 m)	yellow
Yellow bells (*Stenolobium stans*)	15 (4.6 m)	yellow
Northern California		
Blireiana plum (*Prunus blireiana*)	25 (7.6 m)	pink
Japanese apricot (*Prunus mume*)	20 (6.1 m)	pink, white, crimson
Yulan magnolia (*Magnolia denudata*)	40 (12.2 m)	white

George W. Park Seed Co.
Greenwood, S.C. 29647

Seedway Inc.
Hall, N.Y. 14463

W. Atlee Burpee Seed Co.
Warminster, Pa. 18974

Stokes Seeds Inc.
2046 Stokes Bldg.
Buffalo, N.Y. 14240

Earl May Seed and Nursery Co.
Shenandoah, Iowa 51603

George J. Ball
West Chicago, Ill. 60185

*FLOWERS
AND
VEGETABLES*

Kelly Brothers Nurseries Inc.
Dansville, N.Y. 14437

Musser Forests Inc.
Indiana, Pa. 15701

Stark Brothers Nurseries
Louisiana, Mo. 63353

Stern's Nurseries
Geneva, N.Y. 14456

J. E. Miller Nurseries
926 West Lake Road
Canandaigua, N.Y. 14424

Spring Hill Nurseries
110 West Elm Street
Tipp City, Ohio 45371

*NURSERY
MATERIAL
TREES
SHRUBS
PERENNIALS*

A. M. Leonard and Son Inc.
P. O. Box 816
Piqua, Ohio 45356

Al Saffer
130 West 28th Street
New York, N.Y. 10001

S. S. Skidelsky
685 Grand Avenue
Ridgefield, N.J. 07657

Nasco
Fort Atkinson, Wisc. 53538

*HORTICULTURAL
SUPPLIES*

Appendix W

Listing of Poisonous Plants *

Fruits and Vegetables	Cherry Peach Rhubarb	
Trees and Shrubs	Azalea Black locust Boxwood Daphne Elderberry Golden chain tree Holly	Horse chestnut Lantana Privet Rhododendron Wisteria Yew
Garden Flowers	Bleeding heart Foxglove Iris Lily of the valley	Lupine Monkshood Morning glory Poppy
House Plants	Autumn crocus Caladium Crown of thorns Daffodil Dieffenbachia	Hyacinth Monstera Snowdrop Snow on the mountain
Noncultivated Plants	Blood root Buttercup Jack in the pulpit Mayapple Mistletoe Mushrooms Nettle	Nightshade Poison hemlock Poison ivy Poison sumac Skunk cabbage Water hemlock

1. *Common Poisonous Plants* by John M. Kingsbury. Cornell Information Bulletin No. 104. 50 cents.
2. *Plants That Poison.* A pamphlet from Ciba-Geigy Corporation, Ardsley, N.Y. 10502. Free as a public service.

* Some part of each plant listed is toxic to humans, usually when consumed. For additional information, one may consult either or both of the listed publications.

ANNUALS
Low Growers

Annual phlox
Blue torenia
Cape marigold
Dwarf candytuft
Dwarf nasturtium

Iceplant
Pansy
Portulaca
Alyssum
Verbena

Medium-to-tall Growers

Bachelor's button
California poppy
Cosmos
Larkspur
Petunia

Phlox
Poppy
Salvia
Snapdragon
Zinnia

PERENNIALS
Low Growers

Alyssum
Anemone
Arabis
Candytuft
Coral bells
Dwarf iris
Forget-me-not
Fringed bleeding heart

Lily of the valley
Mother of thyme
Primrose
Sedum
Sempervivum
Sweet william
Viola

Medium-to-tall Growers

Aster
Bleeding heart
Campanula
Columbine
Dusty miller
Ferns
Hardy phlox

Lupine
Larkspur
Poppy
Shasta daisy
Veronica
Viola
Virginia bluebells

VINES

Boston ivy
English ivy
Euonymus radicans
Hall's honeysuckle
Myrtle

SHRUBS

Anthony waterer spirea
Deutzia
Spirea frobelia

BULBS	Crocus
	Grape hyacinth
	Lily
	Narcissus
	Snowdrop
EVERGREENS	Arborvitae—globe, dwarf
	Juniper—horizontal (variety)
	Juniper—Pfitzer's
	Yew—dwarf varieties

Shade-tolerant Plants

Acer (some maple species)   Liquidambar (sweet gum)   *Trees*
Cladrastis (yellowwood)   Lirodendron (tulip tree)
Elaegnus (Russian olive)   Quercus (oak species)
Fagus (beech)   Tilia (littleleaf linden)
Ginkgo (maidenhair tree)

Buxus (common and Korean boxwood)   *Shrubs*
Ilex (Japanese holly)
Euonymus (spreading euonymus)
Kalmia (mountain laurel)
Rhododendron (rosebay and others)
Mahonia (Oregon hollygrape)
Pieris (Japanese andromeda)
Taxus (Japanese yew)
Thuja (arborvitae)
Tsuga (hemlock)

Ajuga (carpet bugle)   Lonicera (Hall's honeysuckle)   *Ground Covers*
Euonymus (wintercreeper)   Pachysandra (Japanese spurge)
Hedera helix (English ivy)   Vinca minor (periwinkle)

Festuca (fescues)   *Grasses*
    F. arundinacea (tall fescue)
    F. rubra (red fescue)
    F. rubra v. commutata (Chewing's fescue)
Poa trivialis (rough bluegrass)

Fruits: Pollination Requirements and Planting Distances

Apples. Plant two different varieties.

Blackberries. Self-fruitful.

Blueberries. Plant two or more varieties.

Cherries (sweet). Plant at least two different varieties (Winsor plus another variety). Montmorency and North Star, are self-fruitful. (Sour). Can not be used to pollinate sweet cherries. They are self-fruitful.

Currants. Self-fruitful.

Elderberries. Plant two different varieties.

Grapes. Self-fruitful, except the variety Brighton.

Gooseberries. Self-fruitful.

Peaches. Most varieties are self-fruitful.

Pears. Plant two different varieties.

Plums. Plant two different varieties to ensure pollination. Varieties Stanley and Fellemburg are self-fruitful. The Japanese varieties are unfruitful.

Raspberries. Self-fruitful.

Strawberries. Self-fruitful.

PLANTING DISTANCE BETWEEN SIMILAR FRUITS, IN FEET

Fruit	Standard Size	Dwarf Size
Apple	35 (10.7 m)	10 (3.1 m)
Blackberries	6 (1.83 m)	
Blueberries	3–6 (0.92-1.83 m)	
Cherries, sour	20 (6.1 m)	
sweet	25 (7.6 m)	
Currants	4 (1.22 m)	
Grapes	8 (2.44 m)	
Peaches	20 (6.1 m)	10 (3.1 m)
Pears	20 (6.1 m)	10 (3.1 m)
Plums	20 (6.1 m)	10 (3.1 m)
Raspberries	3–6 (0.92-1.83 m)	
Strawberries	1–2 (0.31-0.61 m)	

Necessary Florist's Supplies and Flower of the Month

Containers—Baskets, Vases
1. Brass
2. Copper
3. Glass
4. Pewter
5. Plastic
6. Pottery
7. Silver
8. Wood

Flowers
1. Fresh cut
2. Artificial
3. Dried

Greens
1. Evergreen—both needled and broad-leaf types
2. Deciduous plants in season

Holders
1. Artist's clay
2. Needle point
3. Oasis
4. Plastic tubes
5. Shredded styrofoam
6. Snowpack
7. Sticky tape
8. Vermiculite

Jackknife and Hand Pruners

Refrigerators
1. For display
2. For storage

NECESSARY SUPPLIES FOR FLORISTS

Ribbons

Tape

Truck—for delivery

Wire

FLOWER OF
THE MONTH

January	Carnation
February	Violet
March	Jonquil
April	Sweet pea
May	Lily of the valley
June	Rose
July	Larkspur
August	Gladiolus
September	Aster
October	Calendula
November	Chrysanthemum
December	Narcissus

Greenhouse Shading and Lighting Requirements

Methods used to shade greenhouses during the summer

1. Whitewash or hydrated lime mixed with water and applied either with a roller or sprayed on the glass. Because of the chemical reaction, this material is not recommended for aluminum-framed greenhouses.
2. One gallon of white latex paint mixed in eight gallons of water is inexpensive, easy to apply with a roller, and should be completely washed off by fall.
3. Commercial roll-up shades constructed from aluminum, Saran, or vinyl plastic provide the grower with the ability to shade plants when needed and to raise the shades whenever the light intensity decreases.

Requirements for increasing photoperiod*

1. Use 60-watt incandescent lamps positioned in clean reflectors.
2. Position the lamps at intervals of 4 ft (1.22 m) and at least 3 ft (0.92 m) above the plant tops.
3. The light intensity at the top of each plant should be at least 10 footcandles during the entire lighted period.

Potted plants that prefer an acid pH between 5 and 5.5

Azalea
Camellia
Cineraria
Cyclamen
Daffodil
Ferns
Gardenia
Hydrangea (blue)
Lily
Orchid

*For a greenhouse bench 4 ft (1.22 m) wide.

Cleaning and sterilizing used pots

Completely submerge the containers for 30 minutes in a mixture of one part household bleach to nine parts of water. The excess soil should be scraped off the containers before they are placed in the 10 percent bleach solution. The containers can be reused the day after they are removed from the mixture.

Greenhouse Manufacturers and Fiberglass Suppliers

Arrow Industries
230 Fifth Avenue
New York, N.Y. 10001

Greenhouseman
P.O. Box 2666
Santa Cruz, Calif. 95060

Lord and Burnham
2 Main Street
Ivington, N.Y.

McGregor Greenhouses
P.O. Box 36
Santa Cruz, Calif. 95063

National Greenhouse Co.
P.O. Box 100
Pana, Ill. 62557

J. A. Nearing Co.
10788 Tucker Street
Beltsville, Md. 20705

Oehmsen Plastic Greenhouse Mfg.
50 Carlough Road
Bohemia, N.Y. 11716

Redwood Domes
Aptos, Calif. 95003

Sturdi-built Mfg. Co.
11304 S.W. Boones Ferry Road
Portland, Ore. 97219

Turner Greenhouses
P.O. Box 1260
Highway 117 South
Goldsboro, N.C. 27530

Plexiglas
115 Cedar Street
New Rochelle, N.Y.

Price Brothers
Highway 57
Grand Junction, Tenn. 38039

Construction of a Cold Frame

CONSTRUCTION
MATERIALS

Wood. Cypress is the best; it resists decay.

Cinder blocks. More permanent. Their original cost is higher than wood. They must be placed below the frost line (3 ft deep) and must be cemented together.

DESIGN

1. The *back* should be 6–9 in. higher than the front.
2. The *front* may vary in height from 6–12 in.
3. The *cross-ties* run from front to back and are spaced every three feet. They provide support for the sash. They must be dovetailed into both the front and back walls and notched on both sides to keep the sash from slipping.

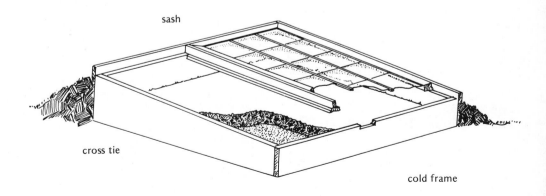

sash

cross tie

cold frame

FIGURE A-15
Construction of a cold frame.

4. The soil depth should be a minimum of 6–8 in.
5. Below the soil level there should be 3 in. of drainage material.
6. The standard size is 3 ft by 6 ft.
7. The sash are usually 34 in. wide and 36 in. long. They may be constructed of cypress, redwood, and Douglas fir. Glass panes are used in the sash.
8. The first glass pane is placed in the rear of the sash, and each additional pane is slid ½ in. under the previous one. Each pane is secured in place by using glass clips and a glazing material.

1. Excessive dampness. Requires less watering and some air movement.
2. Ventilation. Must be done on hot sunny days.
3. Watering. Best accomplished during the early morning hours.

*ENVIRONMENTAL
CONCERNS*

Appendix EE

Basic Greenhouse Construction

Roof rafters. 2 × 4's placed on 16-in. (40-cm) centers. (The distance may be extended to 24 in. (60 cm) but then the rafters need cross members between them for added strength.) When anchored to the house, a 2 × 6 should be used.

End panels. 2 × 4's on 16-in. centers. If end door is desired, double the 2 × 4's on each side and above the door to prevent sagging.

Side wall(s). 2 × 4's positioned on 2 × 4 plates, and lined up directly beneath the rafters.

Foundation. Must go to a depth of 3 ft (0.92 m) around the entire perimeter. It can vary in height above the ground from 8 to 30 in. (20 to 75 cm).

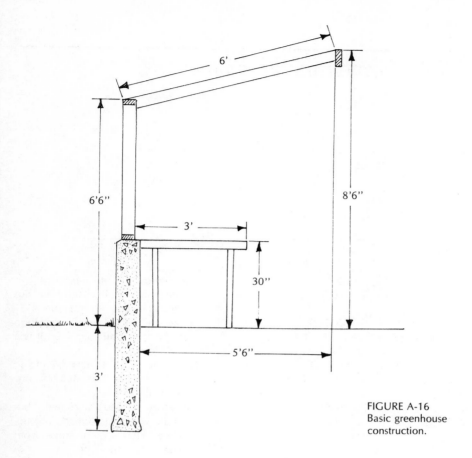

FIGURE A-16
Basic greenhouse
construction.

Appendix FF

Common Vines Grown in the Home

Common name Scientific name	Temperature	Light	Comments
1. Allamanda	60°F (15°C)	sun	Has yellow or purple flowers in the summer. It should be cut back and repotted in January.
2. Clerodendrum (*Clerodendrum Thomsoniae*)	60–70°F (15–21°C)	sun	White, red, or violet flowers form in the spring and summer. It is a fast grower.
3. Creeping fig (*Ficus pumila*)	60–75°F (15–24°C)	semi-sun	Has aerial roots that will cling to any object. The leaves are heart-shaped.
4. Creeping gloxinia (*Asarina erubescens*)	50–60°F (10–15°C)	sun	Has arrow-shaped leaves and tubular flowers that form during the winter. Flower color ranges from blue to rose to white.
5. Devil's ivy (Pothos) (*Scindapsus aureus*)	60–70°F (15–21°C)	semi-sun	Green oval leaves splashed with yellow. The plant is not able to tolerate cold air or water.
6. English ivy (*Hedera . . .*)	60–65°F (15–19°C)	shade	Leaves of different varieties have various shapes and some may be variegated.
7. German ivy (*Senecio mikanioides*)	50–70°F (10–21°C)	semi-sun	Has fragrant yellow flowers and deep green leaves.
8. Grape ivy (*Cissus rhombifolia*)	65–70°F (19–21°C)	semi-sun	Has glossy three-parted leaves.
9. Kangaroo vine (*Cissus antarctica*)	60–75°F (15–24°C)	shade	A rapid grower that is free from pests.
10. Kenilworth ivy (*Cymbalaria muralis*)	60–75°F (15–24°C)	semi-sun	Has violet flowers with kidney-shaped leaves tinted with red beneath. May be called "mother of thousands."
11. Philodendron (*Philodendron cordatum*)	60–70°F (15–21°C)	semi-sun	Easiest of all the vines to grow. Its shiny red buds uncurl into heart-shaped leaves.

Common name Scientific name	Temperature	Light	Comments
12. Ragwort vine, or "little pickles" (*Othonna capensis*)	60–70°F (15–21°C)	semi-sun	Has flowers during the winter and small pickle-shaped fruit.
13. String of hearts, or "rosary vine" (*Ceropegia woodii*)	60–70°F (15–21°C)	semi-sun	Has urn-shaped flowers and heart-shaped leaves.
14. String of pearls (*Senecio rowleyanus*)	60–70°F (15–21°C)	semi-sun	Has small flowers that form during the winter.
15. Swedish ivy (*Plectranthus australis*)	60–75°F (15–24°C)	semi-sun	Popular house plant that has either all green leaves or variegated ones.
16. Wandering Jew (*Tradescantia . . .*)	60–80°F (15–27°C)	sun	A large and varied family whose leaves usually are two-colored; they have weak stems and do not flower.
17. Wax plant (*Hoya carnosa*)	50–65°F (10–19°C)	sun	Has waxy leaves and clusters of small waxy pink flowers.

Appendix GG

Vegetable Planting Information

Vegetable seed	per 25 ft	Planting depth	Distance between plants (inches)	rows
Bean	1/8 oz	2 in.	2–3	18–24
Beet	1/4 oz	lightly cover	sow thin	18–24
Broccoli	1/16 oz		24	24
Brussels sprouts	1/16 oz		24	18–24
Cabbage	1/16 oz		15–18	24
Cauliflower	1/16 oz		24	18–24
Cantaloupe	1/8 oz		48H	60
Carrot	1/16 oz	1/2 in.	1–3	18
Corn	1/2 oz		10	24–30
Cucumber	1/8 oz		48H	48
Eggplant	1/16 oz		30	36
Lettuce, leaf	1/8 oz	1/4 in.	sow thin	18
Onion sets	1/2 lb		3–4	18–24
Parsnip	1/8 oz		4–6	15–18
Pea	1/4 lb	1/2 in.	2	36
Pepper	1/16 oz		20	30
Potatoes, seed	1 bag	8 in.	12	18–24
Pumpkin	1/8 oy		60H	60
Radish	1/4 oz	lightly cover	sow thin	12
Spinach	1/4 oz		sow thin	15–18
Squash, summer	1/8 oz		36H	48
Squash, winter	1/4 oz		60H	96
Tomato	10–12 seeds		24	36

[1]Metric conversions
  1 oz = 28 g
  1 lb = 0.45 kg (454 g)
  1 in. = 2.5 cm
  1 ft = 0.3 m

[2]Other notations
  H plant in hills
  (*) can be started indoors

[3]Rule of thumb: plant most seeds at a depth three times their diameter (except those specifically mentioned).

Planting time	Days to harvest	Helpful tips
after frost	50–65	
after frost	50–70	
May (*)	80–100	Sow indoors 5–7 weeks early
April 20 (*)	90–130	Sow indoors 4–6 weeks early
April 20 (*)	60–110	Sow indoors 5–7 weeks early
April 20 (*)	75–110	Sow ½ in. deep indoors
May 10 (*)	50–80	Same as cauliflower
2 crops	55–90	(1) After frost; (2) June
May 1	70–80	
May 1 (*)	60–80	6–8 per hill; can sow indoors
June 1 (*)	70–85	Sow ½ in. deep indoors
May	45–50	
May	130–150	
after frost	50–80	
Feb. or March	50–80	
May 15 (*)	70–90	Sow ½ in. deep indoors (jiffy pots)
April 20		Should be dusted with sulfur
May 15	90–110	5–6 per hill; can start indoors
after frost	25–35	
Feb.–March	40–60	
May (*)	50–80	3–4 per hill; can sow indoors
May	90–110	
May (*)	70–100	Sow indoors 5–7 weeks early; set plants 5–6 in. below soil line

Hardy Vines that Climb

By twining stems—must be supported Scientific name (Common name)	Light desires	Flowers	Other Comments
Actinidia chinensis (Chinese gooseberry)	sun or shade		Has separate male and female plants.
Akebia quinata (Five-leaf Akebia)	sun or shade	purple	Purple, sausage-shaped fruit, vigorous yet dainty.
Aristolochia durior (Dutchman's pipe)	sun or partial shade	pipe-shaped	Fast grower with large leaves.
Celastrus (various) (Bittersweet)	shade or sun	yes	Red-yellow berries, separate sexes.
Lonicera (various) (Honeysuckle)	sun or partial shade		Hall's variety is the fastest grower.
Polygonum aubertii (Silver fleece vine)	shade	yes	Flowers last a long time; a fast grower.
Pueraria lobata (Kudzu vine)	full sun		Has large leaves; a very fast grower.
Wisteria sinensis (Chinese wisteria) *W. floribunda* (Japanese wisteria)	sun or partial shade	blue or white	Can be grown as a bush or small tree, but will climb to any height.
By coiling tendrils—must be supported			
Ampelopsis brevipedun- culata (Porcelain berry)	sun or shade		Berries change from green through yellow and white to blue; leaves are lobed; not for dry areas.
Bignonia capreolata (Cross vine)	full sun	orange-red trumpet-shaped	An evergreen vine.
Clematis (various)	sun or light shade	various	A very showy vine.
Passiflora (various) (Passion flower)	full sun	blue or white	Has an edible fruit.
Vitis coignetiae (Glory vine or grape vine)	sun		Has brilliant fall color; rapid grower.
By clinging rootlets—some of these require support, while others do not			
Campsis radicans (Trumpet creeper)	sun	orange-scarlet	Has dense foliage with compound leaves.
Euonymus fortunei radicans (Wintercreeper)	sun or shade		Evergreen which is slow to become established.
Hedera helix (English ivy)	some shade		This vine has many varieties; baltica is the most hardy.
Hydrangea petiolaris (Climbing hydrangea)	sun or partial shade	cream colored	Slow to become established.
Parthenocissus quinquefolia (Virginia creeper)	sun to deep shade		Scarlet fall foliage, blue-black fruit.
Parthenocissus tricuspidata (Boston ivy)	sun to partial shade		Strong, rapid grower with large leaves and brilliant fall color.

Aeration.   Supplying oxygen to the roots of plants and microorganisms in the soil.

Aerial root.   Used to absorb moisture from the air. Example: orchid.

Air layering.   Vegetative propagation method whereby the stem of the plant is ringed, packed with moss, enclosed in plastic and left until roots appear. Example: dumb cane.

Alpine plant.   A plant suitable for rock gardens or one that normally grows in exposed mountainous regions.

Alternate.   Term used to describe the arrangement of the leaves and buds on a stem when they occur singly at different heights on the stem.

Annual.   A plant that completes its entire life cycle, from seed to fruit, in one year. It then dies, but some may reseed for the following year.

Annual ring.   The layer of wood laid down by the cambium layer during a single growing season.

Anther.   The portion of the male flower that bears the pollen.

Aphid.   Very small insects commonly called plant lice.

Aquatic plant.   Any plant that grows in or near the water.

Axillary bud.   A bud originating in an axil or point where a leaf stalk or branch forms an angle with the main stem.

Balled and burlapped.   Any plant dug with a soil ball and tied up in burlap.

Bare root.   Rooted without soil, opposite to balled and burlapped.

Bark.   A term used to include all the plant tissue outside the cambium layer.

Biennial.   Any plant requiring two seasons to complete its life cycle, the first for vegetative growth and the second for both flowering and fruiting.

Bract.   A small, reduced leaf found in close proximity to a flower. Example: poinsettia.

Budding.   A specialized type of grafting whereby the bud (scion) is inserted into the stem of another plant (stock).

Bulb.   An underground storage stem containing a flower bud and food-filled leaves.

Burning.   The browning of leaves caused by too much fertilizer.

Callus.   A new covering of protective tissue at the base of a cutting.

327

Calyx. The outer parts of the flower, composed of sepals.

Cambium. A narrow layer of tissues that carries the food manufactured in the leaves to the roots and which produces cells that become bark on the outside and wood on the inside.

Canker. A dead area, usually on a stem, caused by a fungal disease.

Capsule. A seed pod. Example: poppy.

Catkin. A hanging cluster of petalless flowers. Example: birch.

Chelated iron. A guaranteed form of iron that will remain in soluble form because the iron is held in claw (chelated) form.

Chlorophyll. Green pigment found in plants.

Chlorosis. A yellowing of certain plant leaves due to a lack of iron within the soil. Corrected by addition of chelated iron.

Cold frame. An unheated, bottomless structure with removal window sash which are held in place by the frame. It is used to extend the outdoor growing season.

Compost. Decomposed organic matter.

Compound leaf. A leaf composed of two or more leaflets. Example: most ferns.

Concrete. A mixture of cement, sand, and gravel that is used for constructional purposes.

Conifer. A cone-bearing plant, or gymnosperm.

Contact poison. An insecticide that kills on contact.

Corm. A solid, underground stem. Example: crocus.

Cotyledon. An embryonic seed leaf that is easily seen in large seeds. Example: bean.

Cover crop. Plants grown during the dormant season for the purposes of controlling weeds, preventing erosion, and improving the soil.

Crotch. Any angle formed between two woody branches, or that point where the main trunk divides into branches.

Cultivation. The scratching of the soil surface to either suppress the growth of weeds or encourage plant growth.

Cuttings. Separated plant parts that are placed in a rooting medium with the purpose of increasing the number of plants.

Damping-off. A seedling disorder caused by various fungi.

Debris. Waste material that has accumulated.

Decay. Rotting of plant parts caused by fungi and bacteria.

Deciduous. The losing of leaves during the fall by nonevergreen plants.

Defoliant. Any chemical applied to a plant with the purpose of having its leaves drop off.

Dicotyledon. Seed having two embryonic leaves.

Dioecious. Having male and female flowers on separate plants of the same species.

Disbudding. The removal of buds from a plant to stimulate the production of other buds or to increase the size of the remaining buds.

Division. The splitting of a herbaceous plant into two or more sections as a means of increasing the number of plants.

Dormant. Period of time when a plant is resting or not in active growth.

Double. Terms used to describe flowers having more than the usual number of petals.

Downy. Term used to describe soft hairs; also called pubescent. Example: African violet.

Drill. Another term used instead of a vegetable row.

Drought. A lack of moisture that usually hinders plant growth.

Drupe. A fruit with a thin skin, fleshy middle, and stony case surrounding the fruit. Example: cherry.

Drying-off. The process of preparing bulbs and tubers for their necessary rest or dormant period by withholding water.

Epiphyte. An air plant whose leaves are capable of catching water.

Espalier. Method used to train plant material in lattice fashion.

Evergreen. A plant that remains green the year round.

Eye. The growing bud on a tuber.

Fertilization. The union of a male cell with an egg or a female cell to form the first cell of a new plant.

Fertilizer. Any material containing minerals essential for plant growth.

Filament. The stalk of a stamen or male flower part.

Flat. A shallow box used for the germination of seeds or the rooting of cuttings.

Foliar. Having to do with a leaf or leaves.

Forcing. Any procedure that is used to speed plants to grow, or to produce flowers or fruit at times other than normal.

Foundation planting. Those plants found along the base of a building that hide the foundation.

Frond. A fern leaf.

Fruit. The seed-bearing part of any plant, a mature ovary containing seeds.

Fungi. Plants, such as mushrooms, toadstools, mildews, and molds, that lack the pigment chlorophyll.

Fungicide. Any material that destroys or helps to prevent the growth of fungi.

Generation. A single complete life cycle in the life of a plant or insect.

Genus. A group of species of plants linked together by specific botanical traits.

Germination. The sprouting of a seed or spore.

Girdling. Cutting a circular ring around the bark of a plant, usually destroying growth above that point.

Glabrous. Smooth surfaced (not hairy).

Glaucous. Covered with a white powdery material that is easily rubbed off. Example: grapes.

Grading. Leveling off any ground surface for proper water runoff. Procedure used to separate grades of cut flowers.

Grafting. The joining of two plant parts or two cambium areas so that their tissues will unite and grow as one.

Growing point. Found at the tips of both stems and roots where cell division occurs and new cells are formed.

Growth inhibitor. Any chemical that suppresses plant growth. Example: B-nine is used commercially to keep pot mums short.

Growth stimulant.   Any chemical that encourages plant growth.

Hardening-off.   Treatment whereby plants are gradually accustomed to their new environment, which usually has greater fluctuations in temperature and water than their former growing area.

Hardiness.   The ability of a plant to survive in a specific location without protection.

Hardwood cutting.   Cuttings taken during the fall months after the plant has become leafless and dormant.

Heading back.   A basic pruning cut consisting of cutting back the terminal portion of a branch to a bud. It usually causes the stimulation of lateral growth, resulting in a bushy, compact plant.

Heartwood.   The dark wood at the center of a woody stem.

Heaving.   Occurs during alternate periods of freezing and thawing, where roots become exposed and frequently die.

Heavy soil.   A soil predominantly composed of clay.

Hedge.   A row of shrubs or trees used as a screen or windbreak.

Herb.   A soft herbaceous plant frequently used for flavoring, for its fragrance, and for medicinal purposes.

Herbaceous.   Soft, not woody.

Hill.   A circular area housing several vegetable seeds. Example: squash. (The area is flat and *not* raised.)

Hormone.   A growth-regulating chemical produced by a living organism.

Humus.   Decomposed organic matter.

Hybrid.   Any living organism that is a cross between two parents which display different characteristics.

Hydroponics.   The growth of plants by nutrient culture (soil-less).

Internode.   The length of space on a plant stem between two adjacent nodes or stem joints.

Insecticide.   Any substance that kills insects.

Irrigation.   Method of supplying water by artificial means.

Landscaping.   Term referring to all forms of garden design and construction.

Lawn.   Any area planted with grass and mowed consistently.

Layering.   Vegetative method of propagation whereby the unsevered stem forms roots at a point where it is injured slightly.

Leaching.   The removal of soluble plant nutrients from the soil by the presence of excess water running through it.

Leader.   The main trunk or growing apex of a tree or shrub.

Leaflet.   A segment of a compound leaf.

Light soil.   One composed primarily of sand.

Liners.   Small plants placed in nursery rows or grown for additional time in containers.

Loam.   A mixture of sand, clay, and silt plus humus, all in varying amounts.

Malathion.   An organic phosphorus compound used as an all-purpose insecticide.

Midrib.   The main vein of a leaf.

Mist.   A gentle spray of water over the surface of plants.

Miticide.   A chemical used to kill red spider mites.

Mold.   The apparent fuzzy growth of any fungus disease.

Monocotyledon.   A seed containing only one embryonic leaf.

Monoecious.   Male and female flowers borne on the same plant.

Mulch.   Any material applied to any ground surface for the purpose of reducing water loss and weed growth, and to insulate the ground surface against wide fluctuations in temperature.

Mushroom.   An edible fungus, but not a toadstool.

New growth.   Growth of the current year.

Nitrogen.   A gas comprising 78 percent of the atmosphere. A macroelement needed for normal plant growth.

Node.   A joint where both leaves and buds appear on the stem.

Nursery.   Area where young plants are raised.

Nutrient.   Any material that supplies food necessary for plant growth.

Nymph.   The immature stage of an insect.

Old wood.   Ripened wood that is at least one year old or older.

Osmosis.   The diffusion of a liquid through a semi-permeable membrane.

Ovary.   The bottom, enlarged part of the pistil where the seeds are formed.

Ovule.   The small structure in the ovary containing an egg cell. It becomes a seed once fertilization has become completed.

Parasite.   An organism that lives and derives its food from another plant or animal.

Patio.   An open court area in the garden that is usually covered with concrete, bricks, flagstone, or other materials.

Perennial.   A plant that lives for more than two years.

Perlite.   A white silica derivative that is much lighter than sand and is used as a rooting medium or soil amendment.

Pesticide.   A generalized term used to describe any chemical that destroys any organism that makes gardening difficult.

Petal.   A portion of the corolla. A modified leaf.

Petiole.   The stem of a leaf.

pH.   A number used to express the concentration of hydrogen ions in a soil or solution. A pH of 7 is neutral, while a pH less than 7 is acidic and over 7 is alkaline.

Phosphorus.   A macroelement that encourages root growth.

Photoperiodism.   Response of plants to different daylengths.

Photosynthesis.   Process by which green plants are able to convert carbon dioxide from the air and water from their roots, in the presence of sunlight, into food.

Pinching.   The removal of stem tips of herbaceous plants to halt stem elongation and encourage branching or side growth.

Pistil.   The ovule-producing part of the flower which is composed of stigma, style, and ovary.

Pod.   A fruit that splits open along two sides. Example: legume.

Pollen grain.   Dustlike material in the anthers which gives rise to male sperm cells.

Pollination.   The transfer of pollen grains from a male flower to the stigma of the female flower or pistil.

Pollinator.   A plant selected for its ability to supply abundant pollen essential for the pollination of related varieties. Example: apple trees.

Pome.   A type of fruit characterized by a true fruit enclosed by a fleshy receptacle. Example: pear and apple.

Pot-bound.   A condition where the growing container is not providing the adequate space needed for proper root growth, or a condition brought on by excessive root growth and requiring repotting.

Potting mixture.   Any mixture of soil and additives used for the potting of plants.

Preemergence.   A chemical applied to the lawn in early spring to destroy weed seeds before they germinate. Example: crabgrass killer.

Pricking-off.   The transfer of crowded seedlings from the germination flat to another flat, where they are spaced at definite intervals.

Propagation.   Any method used to increase the number of plants.

Pruning.   The removal of any plant part for a variety of reasons.

Pupa.   The inactive stage following the larva stage in the complete life cycle of an insect requiring four stages.

Radicle.   That part of the seed embryo that develops into a primary root during germination.

Repotting.   The removal of a plant from one container and its transfer to the next-larger-size container.

Rhizome.   A horizontal, fleshy underground stem, which is also called a rootstock. Example: German iris.

Root cap.   A protective mass of cells located at the root tip which help protect the root cells in that area as the root pushes through the soil.

Root hair.   A delicate portion of the root whose chief function is the absorption of water and nutrients from the soil.

Runner.   A trailing stem that has roots and often gives rise to new plants at its nodes. It is also called a *stolon*. Example: strawberry.

Sapwood.   The young, living, light-colored outer annual rings of a tree.

Scale insect.   A small, sucking insect with a hard, waxy shell.

Scarification.   The scratching of a hard seed coat to hasten germination.

Scion.   A plant portion, usually of at least three nodes, that is grafted or inserted into another plant, called the *stock*.

Seed.   A reproductive structure housing the embryo and containing enough stored food to ensure germination.

Seedling.   An immature plant that was started from a seed.

Sepal.   A segment of the calyx, or outermost floral part, which is normally green in color.

Serrate.   Saw-tooth edge on certain leaves.

Shrub.   A woody plant having many stems arising from its root system.

Slip.   A cutting or scion.

Sod.   A section of turf or lawn.

Softwood.   Term used to describe certain evergreen lumber. Example: pine.

Species.   A collection of plants that have at least one common distinctive characteristic.

Spindly.   Characteristic of some plant growth when placed in areas of low light intensity.

Spraying.   The application of any chemical substance to the surface of plants.

Stamen.   The pollen-producing male flower, consisting of an anther and a filament.

Standard.   Term used to describe trees or shrubs having an upright, bare stem of more than one foot in height.

Sterilization.   Any treatment using steam, chemicals, or dry heat to ensure the removal of the ever-present disease organisms (actually, it is called pasteurization).

Stigma.   That part of the pistil that accepts the pollen.

Stomata.   Pores surrounded by guard cells that are responsible for the exchange of gases in the leaves and stems for the process of photosynthesis.

Stratification.   The positioning of difficult-to-germinate seeds in a layer of sand and the keeping of this mixture cold, but above freezing, for two or more months.

Style.   A cylindrical tube of plant tissue that connects the stigma with the ovary.

Subsoil.   That layer of soil directly below the topsoil.

Succulent.   Term used to describe leaves as being soft in texture.

Sucker.   A secondary plant shoot that may arise from a stump or from the plant's root area. It is also a term used to describe the excess growth on a tomato plant.

Sunscald.   The burning of a plant's foliage by the sun.

Syringing.   The use of a high-pressure spray of water to prevent wilting, to control red spider mite infestations, and to encourage growth.

Systemic insecticide.   Any insecticide that is applied to the soil in soluble form and absorbed by the plant's roots and which is designed to kill most of the sucking insects as they ingest the sap of the plant.

Tamp.   To firm the soil or sand around a tree or constructional material.

Taproot.   A primary root that continues to grow deeper into the soil.

Tendril.   A modified stem fround on certain vines which enables them to twine and cling to nonplant objects.

Terminal bud.   Any bud found growing at the tip of a stem.

Terrace.   A raised garden area with level top and sloping sides.

Terrarium.   An enclosed growing chamber having high humidity, but no standing water.

Thatch.   Lawn grass residue left behind after each mowing that finally becomes a threat to the health of the lawn.

Thinning.   This pruning practice encourages longer growth of the remaining branches by providing them with more space to grow.

Thrips.   Small insects having scraping mouth parts which do damage to all plant parts.

Topsoil.   The uppermost layer of the soil, which can be further improved by the addition of organic matter.

Transpiration.   The loss of water vapor from the leaf surface.

Transplant.   To remove a plant from one growing position and plant it in another location.

Tree.   A woody plant characterized by a single trunk.

Trellis.   Any structure capable of supporting the growth of vines or other climbing plants.

Tuber.   A swollen underground storage stem that has eyes (buds). Example: potato.

Turgid.   Condition whereby the leaf is completely filled with water.

2, 4–D.   A herbicide, chemically called dichlorophenoxyacetic acid.

2, 4, 5–T.   A relative herbicide of 2,4–D called trichlorophenoxyacetic acid.

Underground stem.   Any stem that grows beneath the ground that displays nodes and buds. Example: tuber.

Understock.   That part of a graft that already contains the roots upon which the new plant will depend. It is also known as the *scion.*

Variety.   A subdivision of a species.

Vegetable.   Any root, leaf, bud, or fruit that is edible.

Vegetative propagation.   General term to describe any technique used to obtain a new plant by means other than pollination.

Veins.   The conducting tissue within a leaf and other plant parts.

Ventilator.   Opening in a greenhouse to ensure the exchange of air.

Vermiculite.   Expanded mica that is used as a rooting medium and soil conditioner.

Vine.   Any plant capable of growing along the ground or having tendrils that permit it to cling and twine to walls and other support structures.

Virus.   A microscopic particle that depends upon other living cells for its nourishment.

Visible spectrum.   Those light rays from red on one end to violet on the other end, i.e., similar to a rainbow.

Xylem.   The conducting vessels within the plant.